Re-Thinking Science

For Didja in memoriam

Re-Thinking Science

Knowledge and the Public in an Age of Uncertainty

Helga Nowotny
Peter Scott
Michael Gibbons

Polity

Copyright © Helga Nowotny, Peter Scott and Michael Gibbons 2001

The right of Helga Nowotny, Peter Scott and Michael Gibbons to be identified as author of this work has been asserted in accordance with the Copyright, Designs and Patents Act 1988.

First published in 2001 by Polity Press in association with Blackwell Publishers, a Blackwell Publishing Company

Reprinted 2002 (twice), 2004 (twice), 2006

Editorial office:
Polity Press
65 Bridge Street
Cambridge CB2 1UR, UK

Marketing and production:
Blackwell Publishers Ltd
108 Cowley Road
Oxford OX4 1JF, UK

Published in the USA by
Blackwell Publishers Inc.
350 Main Street
Malden, MA 02148, USA

All rights reserved. Except for the quotation of short passages for the purposes of criticism and review, no part of this publication may be reproduced, stored in a retrieval system, or transmitted, in any form or by any means, electronic, mechanical, photocopying, recording or otherwise, without the prior permission of the publisher.

Except in the United States of America, this book is sold subject to the condition that it shall not, by way of trade or otherwise, be lent, re-sold, hired out, or otherwise circulated without the publisher's prior consent in any form of binding or cover other than that in which it is published and without a similar condition including this condition being imposed on the subsequent purchaser.

ISBN 0–7456–2607–6
ISBN 0–7456–2608–4 (pbk)

A catalogue record for this book is available from the British Library and has been applied for from the Library of Congress.

Typeset in 10.5 on 12 pt Sabon
by Kolam Information Services Private Ltd, Pondicherry, India
Printed in Great Britain by Athenæum Press Ltd, Gateshead, Tyne & Wear

This book is printed on acid-free paper

Contents

Preface vii

1 The Transformation of Society 1

2 Beyond Modernity – Breaching the Frontiers 21

3 The Co-Evolution of Society and Science 30

4 The Context Speaks Back 50

5 The Transformation of Knowledge Institutions 66

6 The Role of Universities in Knowledge Production 79

7 How Does Contextualization Happen? 96

8 Weakly Contextualized Knowledge 121

9 Strongly Contextualized Knowledge 131

10 Contextualization of the Middle Range 143

11 From Reliable Knowledge to Socially Robust Knowledge 166

12 The Epistemological Core 179

13 Science Moves into the *Agora* 201

14 Socially Distributed Expertise 215

15 Re-Visioning Science 230

16 Re-Thinking Science is not Science Re-Thought 245

References 263

Index 270

Preface

The aim of this work is to present an account of a dynamic relationship between society and science. It seemed to us that the current array of arguments intended to persuade society to support science did not take sufficient account of the developments that have taken place, whether in society or in research, which are discussed both in the scholarly and policy literatures and in the popular press. Despite the mounting evidence of a much closer, interactive relationship between society and science, current debate still seems to turn on the need, one way or another, to maintain a 'line' to demarcate them. Often, too, there is a presumption that communication flows one way – from science to society – with scant attention paid to describing the transformative effects of any reverse communication. The development of arguments which bring current social realities and research practices into line, we believe, requires not so much a clearer articulation of the current arguments, useful as that may be, as a revisiting of the foundations on which they are based. To this end, we have developed an open, dynamic framework for re-thinking science. It is based upon four conceptual pillars: the nature of Mode-2 society; the contextualization of knowledge in a new public space, called the *agora*; the development of conditions for the production of socially robust knowledge; and the emergence of socially distributed expertise. Our conclusion, briefly stated, is that the closer interaction of science and society signals the emergence of a new kind of science: contextualized, or context-sensitive, science. Of course, this book builds on our previous work, *The New Production of*

Knowledge (Gibbons et al. 1994), particularly in its greater elaboration of the significance of the 'social' in the practice and constitution of science, but familiarity with that work is not essential to understanding the argument developed here.

This volume, as was the last, is the outcome of a collaborative effort, albeit, this time, with a reduced team. Its production has occupied our thoughts for nearly three years and over this period we have had meetings in London, Zurich and Stockholm, during which we read, modified and, not infrequently, discarded drafts that had been prepared in the period between one meeting and another. Following our usual practice we have aimed to produce an integrated text rather than a set of individual essays. Working with this intention in mind renders unfruitful all attempts to identify who has contributed what to the final result. In any event, as we have already indicated, this is not our style. We have decided, in this case, to rotate the authorship from the strict alphabetical listing of our previous writing, but we want to make it clear that the new arrangement reflects nothing more than our decision to do so.

Many individuals have helped us along the way. In particular, we would like to thank: Yehuda Elkana, Camille Limoges, Hans-Jörg Rheinberger, and John Ziman, with whom we have discussed our ideas at various stages of their formulation; Alessandro Maranta and Myriam Spörri, who checked our references and completed the bibliography; and Sarah Cripps, who compiled the index.

We owe a special debt of gratitude to Dan Brändström, Director of the Tercentenary Fund of the Royal Swedish Bank and to Thorsten Nybøhm, Director of the Swedish Council for Higher Education whose organizations together funded the project. We also want to acknowledge the special contribution of Roger Svensson, Director of the Swedish Foundation for International Co-operation in Research and Higher Education, whose role was, ostensibly, to provide us with administrative back-up. In fact, he made substantive contributions to our discussions, making available to us insights from his vast store of practical experience, and constantly prodding us to provide concrete instances of our more abstract speculations. Roger has been our colleague now on two intellectual journeys and we hope he will not desert us should we begin to contemplate a third!

As we have indicated, this book was written in intervals carved out of over-busy schedules. This has demanded sacrifices and support from our family and friends. In particular, to Carlo Rizzuto, Cherill Scott and Gillian Gibbons, we simply want to say that we won't do it again, but expect that you know us too well to believe that!

Sadly, in the midst of our writing, Helga Nowotny's brother, Didja, died after an agonizing illness. Working as we do, intensely, in close proximity, over long periods of time, it is to be expected that our thoughts would be affected by his suffering. We would like to recognize his abiding presence in the composition of these pages by dedicating this book to his memory.

Helga Nowotny
Peter Scott
Michael Gibbons

1

The Transformation of Society

Science has spoken, with growing urgency and conviction, to society for more than half a millennium. Not only has it determined technical processes, economic systems and social structures, it has also shaped our everyday experience of the world, our conscious thoughts and even our unconscious feelings. Science and modernity have become inseparable. In the past half-century society has begun to speak back to science, with equal urgency and conviction. Science has become so pervasive, seemingly so central to the generation of wealth and well-being, that the production of knowledge has become, even more than in the past, a social activity, both highly distributed and radically reflexive. Science has had to come to terms with the consequences of its own success, both potentialities and limitations.

In *The New Production of Knowledge*, changes in the constitution of science and in research practice were attributed to the growing contextualization and socialization of knowledge. One of the characteristics of Mode-2 science, we claimed, was that knowledge was now being generated 'in the context of application', and our book contained frequent references, appeals even, to the 'social'. The implication of our argument was that science could no longer be regarded as an autonomous space clearly demarcated from the 'others' of society, culture and (more arguably) economy. Instead all these domains had become so 'internally' heterogeneous and 'externally' interdependent, even transgressive, that they had ceased to be distinctive and distinguishable (the quotation marks are needed because 'internal' and 'external' are perhaps no longer valid categories). This was hardly a bold claim. Many other writers have argued that heterogeneity and interdependence have always been characteristic of science, certainly in terms of its social constitution, and that even its epistemological and methodological autonomy had always been precariously, and contingently, maintained and had never gone unchallenged. In a recent essay in *Science*, Bruno Latour wrote about the transition from the culture of 'science' to the culture of 'research' in the past 150 years:

> Science is certainty; research is uncertainty. Science is supposed to be cold, straight and detached; research is warm, involving, and risky. Science puts an end to the vagaries of human disputes; research creates controversies. Science produces objectivity by escaping as much as possible from the shackles of ideology, passions and emotions; research feeds on all of those to render objects of inquiry familiar. (Latour 1998: 208–9)

Latour goes on to argue that science and society cannot be separated; they depend on the same foundation. What has changed is their relationship. In traditional society science was 'external'; society was – or could be – hostile to scientific values and methods, and, in turn, scientists saw their task as the benign reconstitution of society according to 'modern' principles which they were largely responsible for determining. In contemporary society, in contrast, science is 'internal'; as a result science and research are no longer terminal or authoritative projects (however distant the terminus of their inquiry or acknowledgement of their authority), but instead, by creating new knowledge, they add fresh elements of uncertainty and instability. A dialectical relationship has been transformed into a collusive one. In the sub-head in another article Latour sums this up as 'a science freed from the politics of doing away with politics' (Latour 1997: 232).

So much is common, and uncontroversial, ground. But, even in this more 'open' description, much of the attention remains focused on science rather than society. The latter impinges on the argument only when it touches the former – for example, when controversies about nuclear power or environmental pollution draw in a wider range of actors whose presence and significance cannot be ignored. The perspective is still mainly that of the scientific community(ies) – its composition more heterogeneous, its values more contested, its methods more diverse and its boundaries more ragged, of course, but still distinguishable from other domains such as culture, economy and society. In other words the relationship is viewed principally from one, still dominant, perspective. Indeed, it is possible to read into this more 'open' description of science ('research' in Latour's term) a restatement of traditional accounts of the scientification of society. Science's success has made the world more complicated and scientists must wrestle with the consequences of this complication. But science is still in charge.

It is more unusual to view this changed relationship from the perspective of society. The transformation of society is regarded as predominantly shaped by scientific and technical change. In other words the socialization of science has been contingent on the scientification of society. There are now extended scientific communities and more urgent socio-scientific controversies because society as a whole has been permeated by science, although it is accepted that in the process the culture of science – autonomist, reductive and self-referential – has been transformed into something different: in Latour's phrase, a culture of research which is more populist, pluralistic and open. The 'social' has been absorbed into the 'scientific'. It follows, therefore, that those other aspects of social transformation that appear initially to have owed less to scientific and technical change, even if subsequently they have helped to shape Latour's culture of research, must be regarded as inherently less significant. As a result, changes in the affective and aesthetic domains, so dominant in our definitions of modernity, have rarely been given prominence in analyses of the changing science–society relationship – except, perhaps, to be dismissed as irritating irruptions of irrationality.

In *The New Production of Knowledge*, despite the importance of the 'social' in its account of Mode 2, wider social transformations went largely unexplored. This may have been excusable in the light of the interminable academic literature on modernization and modernity (and post-modernity). The book was never intended to be an essay in social theory – any more than it was conceived of as a tract on science policy. Only in the chapter on the humanities, because of the need to engage wider cultural themes which made it essential to acknowledge other dimensions of social transformation, and in that on higher education, because massification and democratization mean that universities are no longer so intimately associated with the production of scientific and professional elites or the dissemination of a scientific culture, was there any attempt to explicate 'society'.

In retrospect this avoidance of any substantial discussion of the 'social' was a weakness – in three senses. First, it allowed the argument to be assessed purely in narrowly empirical terms, as a more or less accurate account of recent trends in scientific production. For example, Diana Hicks and Sylvan Katz (Hicks and Katz 1996) used bibliometric data to test claims about the growth of networking and collaboration made in *The New Production of Knowledge*. Revealingly their tentative explanation was that this trend was probably an 'internal' phenomenon, the consequence of the end of institution building and budget growth during the 1970s, rather than an 'external phenomenon', the result of the changing dynamics of research itself (in scientific as well as professional and organizational terms), not to say of the emergence of a new relationship between science and society. Second, it made the argument unclear at crucial points. As a result the book was read by some critics as an endorsement of applied science and an apologia for relativism. For example, Paul David characterized our argument as 'a post-modern vision' in which 'mission-oriented R&D is well on its way to displacing discipline-based scientific practice, and becoming an ubiquitous and institutionally decontextualized activity' (David 1995: 14). John Ziman has offered similar criticisms (Ziman 1996). Third, this avoidance of the wider social picture made it difficult to differentiate our argument from those of others like Latour who readily acknowledge the changed relationship between science and society. That difference may lie not simply, or perhaps especially, in more radical notions of the new articulations between them, but in a more radical vision of society. This is important because whether the idea of contextualized science is perceived as substantially different from earlier ideas of science and, consequently, more threatening to the rigour of scientific method and robustness of scientific practice depends on how this 'context', that is, society, is defined. If the evolution of society is defined in terms of benign continuity, the difference and therefore the threat are less. If it is defined in disruptive and disjunctive terms, they are greatly increased. The argument that will be presented here, at its simplest, can be reduced to the assertions that (to borrow the terminology used in *The New Production of Knowledge*) Mode-2 science has developed in the context of a Mode-2 society; that Mode-2 society has moved beyond the categorizations of modernity into discrete domains such as politics, culture, the market – and, of course, science and society; and, consequently, that under Mode-2 conditions, science and society have become transgressive arenas, co-mingling and subject to the same co-evolutionary trends.

The Growth of Complexity

Certainly there appears to have been a remarkable coincidence between the development of more open systems of knowledge production on the one hand and on the other the growth of complexity in society – and the increase of uncertainty in both. The climax of high modernity with its unshakeable belief in planning (in society) and predictability (in science) is long past, even if the popularity of 'evidence-based' research demonstrates the stubborn survival of the residues of this belief. Gone too is the belief in simple cause–effect relationships often embodying implicit assumptions about their underlying linearity; in their place is an acknowledgement that many – perhaps most – relationships are non-linear and subject to ever changing patterns of unpredictability. A good example is the development of chaos theory in the 1970s and its enthusiastic reception by a wider public previously unfamiliar with the phrase and certainly not able fully to understand its technical details or appreciate its scientific significance. For this wider public, chaos theory was a powerful metaphor which vindicated its long-held belief that not everything was predictable – either in science or government or in daily life. The popularization of chaos theory had a double significance, political and scientific. First, because 'experts' who previously had pretended to know (almost) everything were shown not to know as much as they claimed, the political distance between governors and governed was reduced; traditional hierarchies of deference were eroded. Second, in epistemological terms chaos theory, in its metaphorical much more than its technical aspects, appeared to suggest that the link between determinism and predictability had been broken.

In retrospect the coincidence between the degree of order, control and predictability thought to be found in the physical and in the social and political worlds is remarkable. The search for control and the belief in predictability had guided the project of modernization from the beginning. The Clock, and later the Machine, had become the guiding metaphor and dominant iconography of the political order. At first regarded as the worldly embodiment of a cosmic order, later this political order was reflected in, and also celebrated, the machine-like operation and technocratic efficiency of welfare-state capitalism and liberal democracy. In its smooth and predictable functioning, the process of modernization in the highly industrialized Western countries reached its climax during the quarter century after 1945. Moreover, modernization was no longer attributed to the 'hidden hand' of the market or other apparently impersonal forces; instead it was publicly on display for all to see, a powerful affirmation of man's control over nature and society. Any remaining errors or malfunctioning systems could be rectified by more and better science and more ingenious and detailed social engineering. The future, an open-ended horizon, seemed to promise wealth and health for all who remained true to these underlying principles of order and liberty.

Of course, in its early days at the end of the nineteenth and beginning of the twentieth centuries modernity had been a much more tormented, and ambiguous, phenomenon. In its cultural manifestations, at any rate, it was plagued by doubts about promises of a progress that had yet to materialize for the majority of the population. But with mass consumption succeeding, and complementing, mass production, these fears eased. In Western Europe and North America, at any rate, the years after 1945 produced unprecedented economic growth, full employment and material wealth for a population that quickly became accustomed to its twin role as producers and consumers. Predictability and control became the hallmarks of an accomplished modernization arrogantly characterized by assertions of universalism, openness, rationality and efficiency. Science and technology also became powerful metaphors for the transformation of politics; the latter came to be seen as potentially as efficient, predictable and orderly as the former. For a brief period in the 1960s the social sciences, in their capacity as advisers to the political 'Princes' of the democracies of the West, were swept up by the same unprecedented euphoria and naively came to believe they could emulate the triumphant progress of the natural sciences. This period coincided with the Cold War – although it was far from actually being a coincidence. The enemy of the open society, like disorder generally, had to – and could – be kept outside the realm in which control and predictability had been successfully installed. Cartesian dualisms, not only of mind and body, but of right and wrong, of good and evil, of rational and irrational, and of sharp differentiation between modern and pre-modern, were justified by the bipolar configuration of the Cold War world. The reputation and funding of science flourished, as did that of technology, for strategic reasons, partly because scientific and technological success was seen as a key guarantor of national security and partly because the wider scientific and technological enterprise benefited from the spin-off from military uses.

This exceptional conjunction of order and freedom, which produced a fleeting, and misleading, coincidence between the (assumed) regularity of society and the predictability of a progressive science, was destroyed by two great events. The first was the oil crisis of 1973–5. Unexpected and without previous warning, it brought home the vulnerability of a highly industrialized technological civilization to sudden changes in its political and natural environment. It had both political and cognitive consequences. First, a new confrontational discourse was created within Western societies as the hitherto uncontested primacy of economic growth was questioned in the light of the rapid depletion of natural resources and degradation of the natural environment. An international commission set up through an initiative of Norwegian Prime Minister Brundlandt coined the word 'sustainability'. Limits suddenly appeared – first to economic growth and then, in the wake of environmental protest movements against nuclear power, to the feasibility of unrestrained scientific-technological developments more generally.

The state, until now seen as the embodiment of political modernity and technocratic efficiency, also began to run up against its own limits. Decentralization in political authority and administration came to be regarded as a requirement of good governance, and citizens ceased to be seen as passive recipients of public goods to be distributed or re-distributed according to expert systems. Consumers became individualized, as did their ability (and right) to maximize their individual preferences, which were now defined according to models of economic rationality and of utilitarian welfare functionality. These developments, of course, were not uniform. National variations and different types of welfare states persisted. Although after the oil shock nearly all post-war welfare states started to evolve in a similar 'market' direction, their actual trajectories were determined by their previous histories, and their detailed configurations shaped by specific, and even unique, 'local' value conflicts and organizational and professional structures.

At the same time, the sources of scientific and technological knowledge were reshaped by the processes of internationalization and, more radically still, globalization, largely (but not solely) supported and stimulated by the development of new information and communication technologies. Knowledge production ceased to be the near-monopoly of a handful of Western industrialized countries. The configuration of scientific and technological knowledge in the context of concrete application became at times as important as its primary production. Control over geographically widely diffused networks of a partly 'immaterial' quality inherent to the new technologies became ever more difficult to enforce. Moreover, new materials and new production processes began to affect the production system itself, which now became 'flexible', organized 'just-in-time' and around principles dictated by 'lean' organizations.

As has already been pointed out, the popularity of chaos theory in the mid-1970s – the cognitive analogue of the oil shock perhaps – marked the beginning of the end of the dominance of modelling using linear and incremental analytical tools based on a paradigmatic calculus. The use of models, of course, increased and spread into new fields where modelling is less obviously applicable. But modelling no longer provided complete answers; problems eluded its grasp. Many of the insights of chaos theory were made possible by the same impressive advancements in computer technology that encouraged globalization. Non-linearity became the catchword of the day. The enthusiastic reception of chaos theory can be seen as one of the subtle shifts from a culture that valued homogeneity to one that braces itself to live in a world of heterogeneity. Chaos theory captured the imagination of Western intellectuals and, more widely, of an intelligent public. The claim that a butterfly's wing over the Pacific could give rise to a tornado over Texas appeared to support their instinctive view that dynamics – of all kinds, individual, social, political and scientific – were essentially non-linear. And the once robust epistemological link between determinism and predictability was also undermined.

The second event was the equally unexpected collapse of the Communist regimes and the end of the Cold War fifteen years later in 1989. No political theory had been developed that could help to explain the rapid, and disorderly but initially peaceful, transition from Communism to free-market capitalism. Few had anticipated the internal contradictions, and consequently erosion from within, of the Communist regimes. Indeed theories that emphasized the contrast between stable totalitarian regimes, which for that reason had to be confronted and contained, and unstable authoritarian regimes, which might be ignored and excused, remained popular through the 1980s. The political repercussions of the collapse of Communism were felt in East and West alike, and were greatly magnified by its unexpectedness. In countries as different as South Africa and Israel their effect was generally positive, opening up new possibilities of political movement and social reform. In the West, and especially in the United States, their effect has been more negative despite short-lived talk of the 'End of History'. The loss of an external enemy and the collapse of mentalities firmly grounded in a bipolar world have produced unexpected internal political fragmentation and contestation. But, in both East and West, the overall impact of the collapse of Communism has underlined the unpredictability of politics.

More fundamental consequences could also be observed. Although the Cold War embrace between scientific and military systems had encouraged some on the Left to demonize science and technology (already suspect on environmental and egalitarian grounds), these links had contributed more powerfully to science's sense of solidity, utility and linearity. Politically contested (but only by a minority), science (despite – or perhaps because of – this contestation) seemed cognitively secure. But with the collapse of Communism this powerful source of support, political and cognitive, was lost. The half-century persistence of a bipolar Manichaean world had also sustained support for the social engineering of the post-war welfare state. However uncongenial to free-market capitalism and however unpromising as a tool of social-democratic reform, welfare states seemed the price that must be paid to maintain social peace and to ensure the loyalty of the working class. Full-employment policies, therefore, were rooted in Cold War political necessities as well as Keynesian economic theories. The forty-four years of armed peace not only stimulated scientific advance; they also fuelled economic growth. Before the spectre of inflation returned as a result of the oil shock and of the United States' reluctance to raise taxes to finance what was initially seen as a local – and short – war in Vietnam, the economic impact of military expenditure, and its civilian spin-offs, had generally been regarded as a stimulus to growth as well as innovation, rather than as a distortion of the economy. Finally, the frightening certainties of the Cold War perhaps induced a cognitive security that was reflected in the intellectual regularities of that period.

The correspondences between the evolution of social and political contexts on the one hand and intellectual cultures on the other are too suggestive to have been merely accidental. The controlling imperatives of post-war welfare states and of pre-oil-shock economies in the West and the success of science, not only in terms of its political prestige but of its cognitive regularity, are too closely aligned. So the end of the Cold War, even more than the oil shock, represented a radical challenge not only to the political (and social) order that had prevailed in the West since 1945, a period which in retrospect can be seen as an age of equilibrium, although its normative stability was disguised by its technical dynamism. Few people recognized that its contestations at their sharpest, in the 1960s, were in reality contained within stable normative structures – or, indeed, that it was the stability of these structures that had permitted these contestations to emerge. The end of the Cold War was also a challenge to the scientific order which both mirrored this wider socio-political environment and, of course, contributed so powerfully to its technical dynamism and created its most eloquent discourses of legitimization.

The so-called post-modern condition is as much a reflection of these external circumstances as a manifestation of the internal dynamics of the disciplinary cultures of higher education and of science or of the rise of a febrile new intellectual culture closely associated with the late twentieth-century development of the culture industries and, in particular, of the mass media. The rise of post-modernism, therefore, represents a crisis both of social legitimization and of methodological, epistemological and even normative authority – although some would prefer to talk of opportunity rather than crisis. The post-modern condition's cultural manifestations and expressions have been widely described and explored. Not only has the received canon of knowledge been questioned; increasingly the limits inherent to scientific knowledge and the knowable have also been probed. It is now recognized that what can be observed and analysed today is only a momentary view of a long-term process. The temporal dimension of evolution raises the question about the evolution of evolution, including the sources of our own evolution as biological and social beings and the evolution of societies. It is in this sense that a Mode-2 society and Mode-2 science are inextricably bound together.

Two Accounts of Social Transformation

In both social and scientific systems, therefore, a regularity that was limited (because the less predictable was relegated to the fringes of both systems) but also generalizable (because it was governed apparently by rational rules and universal laws) has decayed. It has been superseded by an unpredictability that is both unconstrained (because the 'social' is no longer confined to the instrumental-rational arena and 'science' too has burst its positivistic bonds) and particular, even 'local' (because the intensity and pervasiveness of social and scientific change have made both highly sensitive, and therefore susceptible, to 'local' environments). This shift is reflected in competing accounts of social change because Mode-2 society can be conceptualized in two different ways – either as Knowledge Society or as Risk Society, although both labels are much too simplistic. The Information Society, another label which is also often evoked, occupies a middle ground between the two, but this 'Third Way' will not be discussed in detail here; it comprises discourses about the future direction of socio-economic development derived from the political economy of information and communication technologies. While the social-transformation account leading towards the Knowledge Society privileges the changing modes of production, the 'story' of the Risk Society concentrates on those who are affected – consumers and citizens, patients and clients, in short, lay people as well as 'experts' (and, to some extent, dissolves the differences between the 'lay' and the 'expert'). The 'Third Way', or Information Society account of social transformation, seeks to analyse the implications of information and communication technologies for services related to final users (who, by this definition, are already drawn in and, hence, part of the system and its evolving infrastructure).

These competing accounts are based on two different analytical axes. The first, apparently more relevant to definitions of Mode-2 knowledge production, is the scientific-technical-economic, with a heavy emphasis on production. A convenient label, perhaps, is post-industrialism. The second is the socio-cultural, for which terms like post-modernism, or post-Fordism with its more radical and disruptive undertones, may be a better shorthand description. To the extent that writers on the development of science, and on science and technology policies, have thought seriously about the future shape of society they have tended to emphasize the first axis. New technologies, grounded in the achievements of 'basic' science or seen as its correlate, have transformed the conditions for material production. One result has been unprecedented advances in productivity. Another has been the 'customization' of production which has replaced mass manufacture, although it can be argued that this 'customization' is confined to superficial attributes of increasingly homogeneous products and brands. A third perhaps has been the development of 'virtual' products traded in novel ways, for example, on the Internet. New markets, shaped by these technological possibilities, enabled by affluence and shaped by education, have transformed the conditions for social reproduction. The market, in which materialization seems to open the way to individualization, has become as powerful a social signifier as the collectivities of class, race and gender. Again it can be argued that individualization is produced by the eradication of social 'difference' rather than by the liberation of individuality and that, in any case, these older categories still shape, perhaps decisively, access to these markets in material – and symbolic – goods. Finally, society itself, and the institutions and organizations it comprises, are now organized around the availability and manipulation of 'knowledge' (although this 'knowledge' may be imprecisely defined). A grand chain of being is established, beginning with a new science and proceeding by way of technology and markets to a new society. The resemblance to the regularities of traditional research, which we characterized as Mode-1 science, is clear.

The second axis, the socio-cultural, suggests a different account of Mode-2 society. The impact of new technologies, to which it is argued 'basic' science has often made a surprisingly small contribution, is seen as undermining not only industrial-age patterns of employment but also the meaningfulness of their social constructions – personal, in terms of families and perhaps the notions of intimacy and affection that nuclear 'Western' families reflect and promote, and of communities, whether spatial, in terms of urban and rural 'spaces'; social, in terms of shared experience and collective action; political, in terms of economic planning or welfare states; or national, in terms of distinctive 'histories'. New markets are seen as perverse contrivances, the tantalizing source of ephemeral and volatile identities (whether individual, family or community) which must be reinforced by ceaseless but meaningless consumption. Finally the 'Knowledge Society' is regarded as a dystopia – in four separate senses. First, it promotes inequality by reinforcing distinctions between the knowledge-rich and the knowledge-poor. Second, its 'knowledge' is not wisdom or even science but data, the organization of which is technically (and commercially?) rather than culturally determined. Third, through its pervasive knowledge-data it dissolves traditional canons of art, ideas and artefact and also compromises rational discourse. Fourth, and last, it proliferates risks – environmental, ethical and intellectual – without hope of reconciliation. Clearly such a de-stabilizing account of society cannot be reconciled with the cool rationality characteristic of Mode-1 science. It may even be difficult to combine with the eclecticism of Mode-2 knowledge production.

The sharp contrast between these two accounts can be explained in a number of ways. The most obvious is that the former, the scientific-technical-economic, was first generated in the 1970s under more benign economic and stable cultural conditions long before the collapse of Communism and even before the oil shock. Daniel Bell first offered his account of post-industrialism as long ago as 1973 (Bell 1973). This account has not been substantially modified by successive interpretations of scientific-technological and socio-industrial change, although a slight tendency can be observed for more recent accounts to emphasize its radical and disruptive consequences as the impact of new technologies and markets has been more clearly felt (see, for example, Reich 1992 and Kennedy 1993). Bell's account, as modified by Reich or Kennedy, has stood the test of time in the important sense that it offers a discourse still recognizable in many policy pronouncements and futurological predictions. It has also proved to be a resilient account in a second sense; it has survived the shift from welfare-state bureaucracy to enterprise-state markets without radical modification. Hard technological determinism is never far away. But this account now presents a number of difficulties. It still assumes clear demarcations between the spheres of the state, market and culture (and science and technology), and assumes that they are distinctive formations within society represented by clearly differentiated sub-systems.

This first account, therefore, emphasizes the central role played by technology in reshaping industrial processes, employment practices and social patterns. 'Knowledge', defined both as human capital, in terms of highly skilled work-forces, and as theoretical concepts (or, at any rate, systematized data), has become the key resource in determining competitive success in global markets. The socio-cultural consequences of these changes are typically treated as secondary phenomena – to be optimistically or perhaps naively glossed as 'post-materialism', or tolerated as playful consumerism, or (for the 'losers' in the great game of post-industrial restructuring) managed by benign social policies. Bell's second book, half the length of the original, was merely a footnote, even if the worry-word 'contradictions' was used in the title and Bell had already spotted that 'culture' displayed a clear tendency towards syncretism, rather than following the pathways of further functional differentiation (Bell 1976). The first account reflects the spirit of its age, the third quarter of the twentieth century – its dynamism, in the economic and technological arenas; its social stability and ideological predictability (the two apparently warring blocs of the Cold War era had common roots in rational Enlightenment). As such, it is unreflexively married to a technocratic and presumably also a neo-conservative view of history, social change and transformation.

The second account, the socio-cultural, was generated more recently and reflects the anxiety of its era. The social effects of technology, and of the industrial restructuring its advance allows, are no longer mediated by welfare states and other progressive social policies, because globalization has raised the cost of such mediation, in terms of reduced competitiveness, to an unacceptable – or less acceptable – level. The categories within which such mediation could take place, notably nation-states and cultures, have also been compromised by globalization of arenas and hybridization of environments. The environmental impact of technology and the relentlessly ongoing process of industrialization is also causing increasing concern. This impact is no longer a local phenomenon in the sense of ugly coal mines and belching steel works, urban sprawl and increased traffic, but is now perceived as a general or global phenomenon induced by the creation of polluting mono-cultures to satisfy consumerist cravings and the inability to contain the cumulative effects of otherwise local conditions. Thus, phenomena like global warming and nuclear proliferation exemplify the metamorphosis from local to global and the threats that go with their increasing uncontrollability. The success of science and technology has drawn them into highly contested arenas. Not only has the success of science led to radical modifications of social behaviour (for example, advances in reproductive technology have modified sexual habits and so challenged traditional constructions of intimacy); it now appears to compromise the integrity and uniquencess of human life (as in the examples of gene sequencing and, more recently, cloning). Finally, the end of the Cold War, instead of ushering in an era of ideological stability, as Francis Fukuyama and others triumphantly predicted, has increased the disorder of the world.

Ulrich Beck's argument in *Risk Society* is typical of this second, more disturbing account of future society (Beck 1992). He argues that 'just as modernisation dissolved the structure of feudal society in the nineteenth century and produced the industrial society, modernisation today is dissolving industrial society and another modernity is coming into being' – in other words, industrial society and modernity have become antagonistic in a way that Bell and his successors do not accept, or even recognize (Beck 1992: 10). In his scenario of the 'normalisation of the abnormal' – that is, the self-made production and uncontrollable diffusion of risks – the dominant logic of the industrial age, namely that it can control the risks it produces, is breaking down in an irreversible way. As a result, our 'social mapping' no longer works because we can only conceive of modernity in the context of industrial society. But Beck is not very interested in Bell-type socio-economic analysis. To the extent that he is interested, he offers pessimistic interpretations – for example, by arguing that mass unemployment has been integrated into the occupational structure as 'pluralised underemployment'.

What Beck is more interested in are the social effects of Risk Society. He argues that the modernization of gender roles – more women at work, higher divorce rates and so on – has undermined notions of family, parenthood, sexuality and love which were characteristic of industrial society. As a result, 'The system of co-ordinates in which life and thinking are fastened in industrial modernity – the axes of gender, family and occupation, the belief in science and progress – begins to shake, and a new twilight of opportunities and hazards comes into existence – the contours of the risk society' (p. 27). He then considers the intellectual implications of Risk Society. 'On the one hand, science and thus methodical scepticism are institutionalised in industrial society. On the other hand, this scepticism is (at first) limited to the external, the objects of research, while the foundations and consequences of scientific work remain shielded against internally fomented scepticism. . . . Reflexive modernisation here means that scepticism is extended to the foundations and hazards of scientific work and science is thus both *generalised* and *demystified*' (p. 163). He applies a similar argument to political democracy. Sub-political systems thrive at the expense of grand political structures. Progress and innovation now flow through the channels of business or technology (neither of which are 'democratic' arenas), while democratic institutions atrophy. Anthony Giddens has offered a similar, although less gloom-ridden, analysis of social and cultural change (Giddens 1992).

The contrast between the two accounts of future society is suggestive of both their arguments and the ages in which they were first developed. The first is schematic, linear, confident, while the second is discursive, diffuse and gloomy. The former describes the culmination of past and present trends; the latter their radical subversion. The first emphasizes the primary role of production; the latter, by suggesting that uncontrollable risks have become an integral part of any production process, challenges such a primacy. Consumers, patients and ordinary citizens at the mercy of such a runaway production process are cast into the heroic role of having to resist the self-proclaimed authority of those who still make believe that they know and are in control. The Risk Society is therefore a latent political society, oscillating between public hysteria, tension-ridden indifference and attempts at reform.

Social Change and Knowledge Production

Both accounts of future society can be linked to our account of the transition from Mode-1 science to Mode-2 knowledge production, but in radically different ways. The first suggests a number of phenomena that are consistent with Mode 2. The growth of the 'knowledge' industries has not only led to an increase in 'knowledge' workers and a proliferation of sites of 'knowledge' production, but has also tended to erode the demarcation between traditional 'knowledge' institutions such as universities and research institutes and other kinds of organization. Novel 'knowledge' institutions are arising – in small and medium-sized high-technology companies, for example, or management consultancies and think-tanks (which, arguably, are merely extensions or modifications of traditional institutions; the growth of corporate universities may be a good example). But even more radical change is under way; many, perhaps most, organizations in a Knowledge Society have to become learning organizations, in order to develop their human and intellectual capital, and have also to become increasingly dependent on the 'knowledge' systems to operate efficiently – or at all. In simplistic terms it is even possible to equate the transition from Mode 1 to Mode 2 to the successive step-changes in productivity that have characterized the industrial age and have been produced by the coming together of new technologies, new methods of production (and patterns of consumption) and new energy sources. Why not a fourth – new forms of knowledge production?

However, in some other respects, this first account of social transformation is less congenial to our account of Mode-2 knowledge production. Labour-market statistics do not suggest that the number of 'knowledge' workers is inexorably increasing, certainly not if such workers are narrowly defined as scientific researchers (Sennett 1998). The great growth has instead been in 'data' workers. The ease and instantaneity of communications may even have undermined the need for local sites of knowledge production, even if cost pressures have encouraged out-sourcing of routine 'knowledge' work on a global basis. Global brands, and systems, have flourished – and their proliferation has increased rather than diminished the power of these primary knowledge producers. Nor is there much evidence that the development of a Knowledge Society has weakened the hegemony of traditional 'knowledge' institutions such as universities, although it can be argued that their values and practices have been radically modified by closer encounters with other 'knowledge' organizations not just in government and industry but also in culture (because of the explosive growth of the cultural industries, notably the mass media) and the wider community. The socially distributive and diffusive dimensions of Mode-2 knowledge production are absent. Its potentially transgressive and subversive aspects are limited to cultural syncretism or otherwise ignored. So, although in this first account there are suggestive indicators that support the emergence of Mode 2, there are also counter-trends. The balance-sheet is mixed. Although this account endorses the argument that radical changes have taken place in the organizational structures within which knowledge is produced and its social (and professional) practices, it offers little support for the assertion that core epistemologies and methodologies are also changing.

The second, more radical, account of future society presents other difficulties. Its emphasis on contradictions is difficult to reconcile with the evident continuities implied by the transition from Mode-1 science to Mode-2 knowledge production. As it has been understood, Mode 2 implies an enlargement of the number of participants in research and the widening of what is defined as research. It also implies a multiplication and social diffusion of the sites at which knowledge is produced (rather than the 'abolition' of knowledge – at any rate, knowledge derived from research and embodied in scientific-technical expertise – or its incorporation within the larger categories of data-information and cultural symbols, as this second account suggests). Finally, Mode 2 implies an extension of quality control mechanisms to include new criteria and new constituencies, without denying that demarcations between 'good' and 'bad' research can, and indeed must, still be established. In other words the contextualization of Mode-2 knowledge, implied by the claim that it is produced in a context of application, has its limits. There is still a presumption that rationalizing mechanisms such as politics and the market remain in force to frame this context of application, even if the perspective is widened to include the context of implication. The distinction between the contexts of application and of implication will be discussed in a later chapter.

This second account of future society therefore goes far beyond such open – but still orderly – pluralism as we have espoused. A contextualized 'science' can still cling to its identity, although now enlarged and liberalized, provided the 'context' is sufficiently certain. But the 'context' is now chronically uncertain. Traditional forms of (relatively) stable socialization are being eroded, which begins to problematize all identities; and beyond these unstable forms of socialization are the still more disruptive processes of individualization and globalization within which, arguably, 'science' disappears. Nor does this account suggest that the contradictions it identifies are capable of ultimate resolution. Dialectical relationships, paradigm-shifts, hypothesis–thesis–antithesis are seen as equally anachronistic referential frameworks. Instead this account celebrates different qualities – opacity, fluidity, ambiguity and the self-referentiality of discourse. The affinities to post-modernism are clear. Indeed both this second account of future society and post-modern thought are combined in near-narcissistic notions of 'reflexive modernization' (a concept first developed by Giddens) in which subject and object are no longer clearly distinguished and the 'other' is absorbed into the familiar (which now generates its own contradictions and interrogations). As Beck argues: 'The "objects" of scientisation also become subjects of it, in the sense that they can and must actively manipulate the heterogeneous supply of scientific interpretations' (1992: 156–7). This association with post-modernism is almost too much for our Mode 2. On the one hand, its kaleidoscopic transgressions mirror the eclecticism of Mode 2; on the other hand, it subverts the possibility of generating rational discourses. Arguably, this second account of future society can be said to support the development of novel epistemologies and methodologies (at which the first account balks), but it is indifferent, even irrelevant, to the processes of re-institutionalization and continued re-configurations of various kinds on which our account of Mode 2 also relies.

So both accounts of 'society' create difficulties for our account of Mode 2. Yet these difficulties have to be confronted because of our emphasis on contextualization. Our understanding of the 'social', therefore, has to be clear. The first account, which emphasizes the scientific-technical-economic axis of change, is too linear, too cool, too deterministic. It offers convincing, but limited, support to the idea of Mode-2 knowledge production as a shift in the social practices and institutional bases of research, but almost none to its claim to describe a shift in core scientific values and methods. In contrast, the second account, which emphasizes the socio-cultural dimensions of change, is too fluid, too allusive, too regressive even. It reduces knowledge production to either a post-modern world-game or, in the guise of 'technology', a kind of vicious (as opposed to virtuous) cycle of inquiry in which each step forward creates new and dangerous 'risks' and, in the guise of 'science', a self-referential, even incestuous, activity in which it ceaselessly interrogates itself.

Is it possible to devise a better account of social change that combines, perhaps even reconciles, or transcends both of these accounts and draws on our own analysis of Mode-2 knowledge production? In particular, is it possible to suggest plausible connections between social change and intellectual as well as scientific change – not in terms of 'externalities' such as influences and outputs as described in what some authors see to be the social shaping of science or the scientification of society, but 'internalities', affinities between the two processes of change? Once the traditional distinction between 'external' influences on science and its 'internal', self-propelling dynamics had to be abolished, replaced by a definition of science as a form of culture and therefore constituting a set of cultural practices, the door has been pushed open to allow an exploration of culture in this dual sense. Or, more simply, what are the implications of deeper contextualization for the scientific tradition and research methodologies?

Various attempts have been made to come to terms with different ways in which reflective people try to make sense of the world. For instance, a distinction has been drawn by Richard Rorty between 'solidarity' and 'objectivity'. By 'solidarity' he means reflection on personal, intuitive, group, community experience (which others have described as 'local' or tacit knowledge). By 'objectivity' he means reflection in the context of what he calls 'non-human reality', whether God or scientific truth (Rorty 1991: 21). Seen in such a framework, Mode 2, whether in science or society, may represent an intensification of both forms of knowing, with 'solidarity' reflecting the pressures of contextualization – and so de-differentiation – and 'objectivity' representing the forces of specialized, high-technology, 'expert' society. It is this dual intensification, the merging together of different ways of knowing and their inbuilt rivalry for legitimacy and challenge to the respective monopoly of the other way of knowing, which produces heterogeneity, pluralism and fuzziness. It is probably wrong to conclude that 'solidarity' (interpreted as contextualization) is winning out over 'objectivity' (in the form of 'closed' systems of scientific expertise). But it may be significant that, were this to be the case, there would be much less need for epistemology – and almost none at all for meta-discourse. This might help us to explain our difficulty in penetrating what we will call the epistemological 'core'. It would strengthen the argument that this 'core', if it ever properly existed, had been progressively emptied by both scientific reductionism, because the generic value-structures and methodologies that are left are essentially banal, and by Mode-2 contextualization which undermines all forms of universalism, substantive or methodological. These issues will be discussed in detail in chapter 12.

Here perhaps we are groping towards understanding the possible affinities between broader social change, whether seen in technological or cultural perspectives, and the changes in scientific practice, as characterized in a more focused way in *The New Production of Knowledge*. It may be a beginning to building the links between Mode-2 society and Mode-2 science. Latour's 'culture of research' and Giddens's reflexive modernization, although very different conceptions, are both related to what we have written about the contextualization of knowledge production. The transformation of the science system can be observed not only in changes in its institutional features or modes of funding or a greater shift towards public accountability. There have also been changes going to the very heart of current scientific practices, that are very different in their flows and forms from reductionist science and its attendant epistemologies and methodologies. In reductionist science, still the mode to which many scientists aspire, it is necessary to establish clear boundaries that frame the scientific arena, and distinguish science from non-science. Mode-2 knowledge production, in contrast, takes place within and between open and shifting boundaries. It consists of the reconfiguration of knowledge and people. It is transgressively bounded because, in ways that still need to be spelled out in detail, a new kind of integration with the context is made possible. This integration transcends the channelled and predictable interactions which take place under a socio-scientific regime that still practises (and believes in) some form of separation or segregation.

Another way of putting it is that 'people have been allowed a place in our knowledge' and thus, 'the context (can and does) speak back' (Nowotny 1999: 253). Pre-existing contexts, and deep social substructures, influence science-before-the-event, just as its future impacts anticipate science-after-the event. The setting of priorities and the patterns of funding are not self-evident or self-referential; rather they are the result of complex negotiations in a variety of contexts, where expectations and vested interests, unproven promises and mere potentials play a role. But the context of application which figured so prominently in our previous book needs to be complemented. There is clearly also a context of implication which also needs to be explored, and both are preceded by processes of context formation. The next stage in the argument, therefore, is to see whether the transgressivity of Mode-2 knowledge production is typical of a larger transgression of institutions and categories in contemporary society. If this is the case, it is then necessary to explore whether there are co-evolutionary processes at work both in science and in society.

2

Beyond Modernity – Breaching the Frontiers

The argument in this chapter can be simply stated. Contemporary society is characterized – irreversibly – by pluralism and diversity and also, we argue, volatility and transgressivity. It can no longer be understood either in terms of the norms and practices of scientific rationality, although the force of these norms and practices continues to be felt in daily life, because they have been modified, if not undermined, by Mode-2 knowledge production; or in terms of hegemonic forces such as the market (or 'commodification') to which other perspectives have been subordinated, although we accept that our account of social change can be criticized for paying insufficient attention to power. The great categorizations of the human enterprise produced by the successive revolutions of modernity – scientific, political, cultural, industrial – and around which the contemporary world is organized now appear to be either in flux, eroded or socially contested. It is increasingly difficult to distinguish between the domains of the state and of the market, between culture and mass media, between public and private arenas. Science itself is increasingly challenged by forms of knowledge production, which are subsumed under the term 'research', epitomizing its potential for innovation and exhibiting its seductive lure to politicians and policy-makers alike. Several of these contestations are familiar enough; cumulatively they have contributed to the popularity of post-Fordist accounts of future society and their post-modern analogues, but collectively their mutual affinities have gone largely unexplored.

Reactions to the argument presented in *The New Production of Knowledge* did not help to clarify these connections – either discussion of our account of Mode-2 knowledge production has been contained within the limits of science and technology policies and industrial restructuring, because these wider affinities have not been sufficiently recognized; or our argument has been taken merely to contribute to the development of meta-theories of socio-cultural change, in which new patterns of knowledge production are treated peripherally. In the first case Mode-2 knowledge stands in isolation, because its context of application has not been elucidated in sufficient detail; the 'social' other is assumed rather than analysed. In the second case Mode 2 is equally difficult to comprehend because the role of science is a diminished one – and, in grander theories, demeaned. The aim in this chapter is to offer an intermediate account that emphasizes the affinities between various forms of Mode 2-ishness, not only in research but in politics, the economy and culture.

The State, the quintessential political formation of the modern world, is undergoing a series of profound transformations, in at least five senses. First, nation-states, which can be regarded as Mode-1-like institutions because they seek to embody ideals of bureaucratic rationality (although they also represent darker, more atavistic identities), are being undermined – from 'beneath' by the re-emergence of suppressed local and regional loyalties (whether the 'failed nations' of eighteenth- and nineteenth-century Europe or pre-colonial associations), and from 'above' by the development of supra-national groupings such as the European Union. They are being more radically undermined by the forces of individualization and globalization. In their own chosen terms, those of centralized bureaucratic rationality, nation-states have become dysfunctional – too large to relate to local communities that are also globally constituted and connected and, because of new information technologies and global branding, no longer require mediation between the local and the global; and too small to cope with the emergence of world-economies and world-cultures.

Second, the demarcation between public and private spheres, with the state as guardian of the former and the latter the domain of individual taste and market exchanges, has been eroded. Part of this erosion has been deliberate, such as the privatization of hitherto 'public' services and utilities. But part has been less deliberate and reflects the state's re-definition of its role in 'market' terms, as the purchaser of goods and services. This has taken various, and confusing, forms – the devolution of budgetary responsibilities to agencies that are urged to be more business-like, and the creation of 'internal markets' in which different public agencies are encouraged to trade and compete.

Third, traditional notions of the nation-state have been challenged by the transformation of the political process. Because of its relative powerlessness in the face of globalization, the state's former responsibilities to ensure full employment, to develop progressive social policies, to build science and technology infrastructures have dwindled and the vacuum has been filled by the New Politics of gesture, image, style. As Yaron Ezrahi has put it, state-craft has been replaced by stage-craft (Ezrahi 1990). Democratic politics are no longer a 'production' activity, designed to build centralized bureaucratic structures to promote reform and improvement; they are now a 'consumption' activity, designed to gratify, excite, appease. Symbolic politics has transformed itself from being a mere adjunct to a more 'substantive' core of political activities to occupying its place. As a result the state has taken on many of the characteristics of the market, norms as well as forms.

Fourth – an apparent contradiction – the state has been pushed into new arenas once seen as reserved to the individual or the market. Because of the pervasiveness of the mass media in Western liberal democracies, politicians are required to have views on almost everything – and to take appropriate actions. It is on mishandling of these symbolic-moral issues that governments or individual members now frequently stumble, rather than failure of their large-scale social and economic policies. But there is a compensatory mechanism at work as well. The more constraints on the power-play of 'real' politics there appear to be, the more political activities need to shift to 'mere' expressive gestures. But the terror of transparency is such that where there is no content, the form has to be all the more perfect.

Fifth, the advance of science and technology has also enlarged the territory of the 'political', creating the need for an array of new regulations and regulatory frameworks, notably in the bio-medical area. Although older patterns of closed bureaucratic regulation persist (for example, in the regulation of some drug trials), new kinds of more open regulation are emerging. These regulations are preceded by elaborate negotiations, mediations, consultations and contestations which take place within the public arena – or, as we would prefer to call it, the *agora*. This is no longer the domain of a relatively closed bureaucratic-professional-legal world of regulation, but of broader cultural-political movements embodying antagonistic forms of interaction which have become part of the repertoire of how novel technologies are embedded and research products come to be accepted and used in wider social contexts. As a result of these and other changes, the state has become a transgressive institution, penetrated by but also penetrating market, social movements and individual responses by consumers and citizens.

Something similar has happened to the market. Economy and polity are now confusingly overlaid, as the former increasingly trades in ephemeral images and symbolic goods which are far less amenable to any economizing logic than the durable goods and tangible services of industrial society, and as the latter has also become a not dissimilar 'trading zone' equally unamenable to the claims of bureaucratic rationality. The latter phenomenon, the state as mere mediator or facilitator, is easier to describe. It takes three main forms. The first is the erosion of the public-service ethic (so well described by Max Weber and so crucial for the constitution of the nation-state) that came to dominate the public sphere during the nineteenth and twentieth centuries, and its replacement by a 'business' discourse, still in search of an adequate business ethic as its correlate. As a result hospitals and universities, to take two prominent examples, have increasingly come to be judged not with reference to this declining public-service ethic or their own professional value-structures, but in terms of their 'business' performance (Power 1997: 114–15). This change reflects more than an opportunistic appropriation of a politically favourable discourse which would leave their essential operation relatively untouched, or even an ideological shift (which has been imposed from 'outside'). It also reflects the new 'market' roles now undertaken by health services and higher-education systems. What counts is output and performance, easily measured by performance indicators and similar, seemingly objectified, criteria. Social control no longer needs to rely on personal motivation or professional ethics. Rather, paradoxically, by appearing to 'liberate' the individual from such conscience-based strictures, it actually devolves responsibility because accountability is merely judged in terms of measurable objectives. Another, more realistic or even pessimistic, way of expressing this is to say that there has been a shift from 'social' accountability based on trust and mutuality to 'auditable' accountability based on formal contracts.

The second form in which the state appears as mere facilitator or mediator for the market is the deliberate creation of 'internal markets', and other market-mimicking mechanisms, to resource and manage large parts of the public sector. These systems have tended to replace explicit planning systems. Public accountability is now achieved by auditing their outputs rather than planning their inputs. These systems, sometimes promoted under the name of New Public Management, have also transformed notions of institutional autonomy: normative freedom has often been subtly re-defined as operational discretion. They have also tended to erode the state's public responsibility for its own distinctive domain: 'public' institutions are merely the providers of 'public' services, to be judged in terms of their instrumental efficiency rather than normative significance. Finally, there is the large-scale privatization of public services and utilities, for both opportunistic and principled reasons, which has further eroded the distinction between state and market. As a result it is difficult to say whether the state has shrunk and the market expanded, or the other way round (because the number of private enterprises now 'controlled' by the state, whether through subsidy or regulation – or both – has increased). The reason for this difficulty is that these categorizations have themselves become anachronistic.

The transgression of the market is more difficult to describe. Its rationality as a device for allocating resources, distributing life-chances and generating wealth has always been challenged, although that challenge is weaker now than it has been for more than a century. Furthermore it can be argued that there are no essential differences between the trading of symbolic goods through the mass and electronic media in the late twentieth century and the trading of consumption goods through the complex networks of producers, distributors and consumers that developed in the later nineteenth century. In its time, the department store was as radical an innovation as the Internet in reshaping the world of shopping. Dichotomies between symbolic goods and material goods, between consumer and investment goods, even between goods and services, have always been largely artificial. The case for arguing that the idea of the market, like that of the state, has been transformed in a given direction is not self-evident.

Four strands of evidence can be offered. The first is that the market has become increasingly insubstantial and metaphorical. In many respects it is as much an allegory as an instrument. And, as discourse, it is almost infinitely pliable, leading to a proliferation of different types of markets which are hard to distinguish and define. The 'market' label is now used to describe a wide range of social, political, cultural, even intimate activities far removed from classic market exchanges in identifiable economic arenas. In the process it may have been drained of substantive meaning. Second, these metaphorical-market exchanges have not simply accelerated: they have often become instantaneous. Global brands, ephemeral images, 'virtual' products have become the dominant outputs as well as inputs of the market. The market today is as deeply engaged at the micro-level, with individual feelings and perceptions, as at the macro-level, with the overall allocation of resources and rewards. It is everywhere and nowhere.

Third, partly as a result of this instantaneity, old demarcations between producer, supplier, distributor and user have broken down. Temporal relationships, defined by storage, immediate access and similar shifts, have been transformed. Products can now be quickly re-designed and re-engineered to match consumer preferences, transforming in turn the balance of social and economic relationships characteristic of classic markets. Indeed, the notion of an 'infinite series', originally proclaimed in the manifesto of the US agrarians in the 1930s, seems to have been finally attained: an open-ended progression of invention, learning, adaptation and change (Postrel 1999: 59). Fourth, the market has invaded the domain of intimacy. The commodification of family life and sexual relationships and growth of new technologies of reproduction have reduced the last redoubt of intimacy. There is no longer shame, no restriction of an inner, intimate domain which is kept free from outside intrusion, if it can be marketed, usually with the help of the media. The market is not only a global but has become a personal phenomenon. It no longer has valid boundaries. But this absence of clear demarcations between the market and other domains may mean that classic market forms have themselves been compromised.

Culture has also become a transgressive domain. Once seen as an autonomous space, itself the outcome of a struggle to win autonomy and to insulate itself from market and political demands, culture has now become an integral part of both – in a double sense. First, cultural images have become more significant in political action and market exchanges. The insubstantiality of contemporary politics and the market owes much to their penetration by these images. Rhetorics have superseded policies and programmes. Promises are no longer validated by being fulfilled, or 'kept', in concrete terms; instead they are valued for the images which they conjure up and the potential which they convey. Second, culture has itself been commodified. The 'cultural industries' are among the most dynamic sector in many advanced economies, and the demarcations between high and low culture, between culture, entertainment and information, have been broken down. How new these developments are, however, is open to question. It can be argued that they represent a regression to older forms. For a long time before the emergence of the activist nation-states characteristic of the industrial age, politics was essentially symbolic and, therefore, better analysed in cultural or anthropological terms than in those of modern political science. The medieval (and early modern) belief in the King's ability by his touch to cure scrofula was both a 'political' and a 'cultural' phenomenon. Similarly, our view of culture as an autonomous activity dates back, at the earliest, to the late eighteenth century and was only fully established in the course of the nineteenth. For the sake of brevity, let us say the evolution of this view was coterminous with the life of Goethe.

In his biography of Goethe, Nicholas Boyle describes the early stirrings of this view of culture in his discussion of one of Goethe's early works, the drama *Prometheus*. Jupiter tells Prometheus that he will bring statues to life provided that he, Prometheus, will desist from rebellion against the gods. Tempted Prometheus pauses:

> It is a pause of the deepest significance in the history of modern sensibility. Is the mortal, transient artist the servant of an independent world-order which stretches before and after him, to whose laws he acknowledges himself subject and parts of which are imitated in his works? If so, a realistic, objective, living art is possible, of the kind we associate with Homer or Shakespeare (or perhaps, outside Germany, with the nineteenth-century novel). But if not, if the artist remains an autonomous creator, acknowledging no ordering force except that which he finds within himself, can his work ever escape from its dependence on him? Must it not remain, as in the start of Goethe's drama, stone statues scattered through Prometheus' grove, free but motionless – frozen icons of the artist's self. (Boyle 1991: 165–6)

The whole of art, and culture, can be represented as a meditation on that tension between contextualization and autonomy. The latter view never went uncontested. Its apparent dominance was itself contingent and so contextualized – partly on two revolutions in feeling, the rise of Sensibility in the third quarter of the eighteenth century (the moment of Goethe) and of Modernism on the cusp of the nineteenth and twentieth centuries; and partly on the twin-growths of the bureaucratic state and of a more 'commercial' society, both of which were based on increasing functional differentiation. Although bureaucratized, patronized, professionalized, and commercialized, culture paradoxically came to be seen as the Other, an extra-political and extra-market preserve where holism could be maintained in the face of the splintering of modern industrial life (whether to protect High Culture and traditional values, or to create the space for radical innovation). Historically, these developments are intimately connected with the rise of the bourgeoisie, and the *Bildungsbürgertum* in particular, and the emergence of selfhood as its concomitant and enabling predisposition.

Such a view is now more difficult to maintain. High Culture has been reduced to an elite component within the spreading cultural industries. Traditional values have been further eroded by Beck's and Giddens's 'reflexive modernisation'. In multi-ethnic societies, traditional canons of thought and conduct which once underpinned High Culture have been officially demoted and relegated merely to marginal 'options'. Individual creativity, because creation and dissemination are now intricately combined, no longer flourishes within a segregated territory; instead it has been linked to the process of innovation. The very identity of the artist has become problematical. In a sense we are all artists now. In an age of symbolic exchanges, successful politicians or business people have equally acquired characteristics previously associated with artists. In business and politics, as well as culture, creativity is exhorted as a supreme good; it is 'taught' in management seminars. As a result of this intense syncretism the artistic avant-garde has long since lost its bite. *Epater le bourgeois* has become an increasingly short-lived enterprise creating the briefest of shock waves. Although art still claims to mock and distort, rather than to imitate, wider social trends, nevertheless it increasingly depends on forms and performance, materials and their materialization, borrowed from elsewhere, the world it mocks. Also, by striving so hard to be the 'Other' in a critical and normative sense, art has tended to break away from the professional and institutional structures which guaranteed its operational 'otherness'. Similarly, cultural objects, despite their importance as raw materials in the cultural industries, in museums and tourism, art and entertainment, cannot sensibly be distinguished from the many other symbolic objects that litter the modern and post-modern world. Symbolic objects and symbolic forms of communication have become the staple material of the advertisement industry, from which there is no escape. As a result art arises in everyday life just as powerfully as in designated cultural arenas. Culture is not just contextualized but transgressive. The question is no longer who appropriates whom and what; the process is one of mutual appropriations, leading simultaneously to mutual expropriations.

With the state, market and culture a similar picture emerges – of categorizations that increasingly lack clear boundaries. Perhaps they also make less and less sense, thus fuelling the search for new forms of symbolic expression and leading to post-modernist hyperbole. Just as Mode-2 knowledge production has overlaid and confounded Mode-1 disciplinary science by its heterogeneity and pluralism, so the State, Market and Culture have become 'fuzzy' or blurred categories that defy the orderly picture of functional differentiation under conditions of modernity. Science too has become 'fuzzy', not because of its overall failure to demarcate its boundaries and still less because it is running up against its limits, but because its success has pushed it into ever more contextualized and contextualizing arenas. There the criteria of what constitutes 'success' are contested on grounds of 'values' that defy being opposed to 'facts'. The demarcations between science and non-science are no longer evident, whenever the context of application merges seamlessly with the context of implication. In this sense, the 'limits' of science too are contested. Science is invasive, not invaded. Its autonomous space is no longer guaranteed, since the potential guardians, the state, market and culture, are no longer recognizable there in their old identities, functions and roles. It is in this sense, perhaps, that it is reasonable to speak of the emergence of a Mode-2 society in whose shadow the shift towards Mode-2 knowledge production takes place.

Industrial society, and indeed the process of modernization, has been characterized above all, and has been driven, by processes of functional differentiation, as more and more specialist activities have developed that require specialist institutions embodying conceptual as well as material levels of organization. The state, market and culture, and the relative autonomous spaces they occupied, were products of this differentiation – as was science. The society of the future, while being more specialist still in its technical processes, may be less well differentiated. This de-differentiation, which is an important strand within our analysis of Mode-2 in science and research and also within the wider notion of Mode-2 society, has been greatly enhanced and aided by the rise of increasingly transgressive and instantaneous technologies, techniques and 'boundary objects' that easily cross time and space, and travel from one research site to another (Star and Griesemer 1989: 393). On first view, the pervasiveness and invasiveness of information and communication technologies can be regarded as consistent with the first, and more optimistic (and mechanistic) account of future society, although they deny its underlying linearity and, in fact, undermine the very categories and classifications on which this account is based. But de-differentiation, as highlighted in reflexive modernization and in the second, more sombre, account of social change, is not conducive to institutional solidity. It makes for societal volatility, without 'anchoring' devices. The alternative to accepting that the traditional authority of science has been repudiated and nothing put in its place, and that knowledge is simply replaced by non-knowledge, is to adopt a wider framework and a more long-term view. It is in this regard that our analysis not only of social and cultural change but also of the changes within science, and our argument that they represent a co-evolutionary process, take on both theoretical and empirical significance.

3

The Co-Evolution of Society and Science

In the first two chapters the growth of complexity, unpredictability and irregularity in both science and society was highlighted – and the argument advanced that their co-evolution was more than mere coincidence. We discussed the shift from confident and unproblematical forms of social forecasting fuelled by technological determinism, the Knowledge Society, to much less predictable styles of socio-cultural analysis reflecting the growth of intellectual and social volatility, the Risk Society. The emergence of transgressive institutions, and the blurring of seemingly self-evident categories as the State, Market, Culture took on new and less predictable forms was considered. Finally, we considered the possibility of correspondence between new ways of thinking about science and research, such as we attempted in *The New Production of Knowledge*, and new conceptions of modernity and rationality.

All these coincidences and correspondences between the development of society and science suggest that a process of co-evolution is at work. In the final part of this chapter an attempt will be made to explore in greater detail the possible links between some of these co-evolutionary processes. Viewed in the wider social and political context in which Mode-2 knowledge production emerged, these processes may provide a framework, however crude, for more detailed examination of the dynamics of co-evolution. Although it would be wrong to aspire to identify common patterns of causality, the identification of these co-evolutionary processes may help to advance the argument beyond simple alignment of superficially similar trends in science and society. Viewed from a wider societal perspective, science and technology are deeply implicated in these co-evolutionary processes, but they are far from being the driving forces. Our categories are still deeply entrenched in a dichotomous science–society framework which makes it difficult to increase an understanding of their mutual interdependencies. By adopting an evolutionary framework, largely driven by self-organizing processes, we hope to highlight the parameters that are integral elements in the underlying dynamics.

In the most elementary sense science and technology are implicated above all in an instrumental-utilitarian mode. At the dawn of modern science Francis Bacon's vision was both Utopian and utilitarian. At the turn of the twenty-first century the relationship between science and society is still dominated by interpretations of the contribution science makes towards the enhancement of innovation, wealth creation and economic competitiveness. But utilitarianism's domination of public discourse about science should not detract from a much wider and open-ended spectrum of views of science embracing both the instrumental and the imaginary. Science and technology are valued for their capacity to create new knowledge and deliver an apparently endless stream of new products, but they are also an equally limitless source of new desires and wishes – which can only be satisfied by 'more' science and technology. In this sense science and technology dominate in a double sense – by delivering 'real', or tangible, results; and by creating insatiable images. However, the mechanisms which shape these desires in the first place and then control access to these inexorable processes of interminable wish-fulfilment and regulate their diffusion are social in origin. So too are the forms taken by this wish-fulfilment. In this sense science and technology occupy a subordinate role. In describing the dynamics of the intertwining process between science and society this transient and ambivalent balance between the dominant and subordinate is important.

Transgression is now a major theme. The changing balance between the state and the market, which has already been discussed, is relevant here. On the one hand the state has apparently retreated, reluctantly yielding the roles as protector and patron it had assumed in the era of the welfare state; on the other hand it may have taken on powerful new symbolic roles (or reverted to older nationalist identifications which, in the post-war age of linearity, regularity and rationality, appeared to have become anachronistic). More recently the process of globalization has merely stretched boundaries that had already become highly permeable. Nation-states have remained locked into a system of semi-fictional sovereignty, while the international arena is only sparsely furnished with international institutions that work reasonably efficiently. But global capitalism is subject to no such constraints. In particular since the collapse of Communism, political obstacles and social inhibitions which formerly restricted the application of a crude economic rationality have been cleared away. It began its reign apparently unfettered by other kinds of rationalities.

Today, however, for reasons that have already been explained and which include the emergence of Mode-2 knowledge production, it is not the triumph of one form of rationality (the market) over other forms (such as the social-reform imperatives of the welfare state) that needs to be emphasized, but the erosion of the boundaries between different forms of rationality. The highly specialized and differentiated system of modernity based on a more or less tightly segregated set of different sub-systems, each invested with a specific rationality, an ethos and a set of norms (or, as Luhmann has argued, a specific 'code' as the basis for the crucial distinctions that mark their performance) is now dissolving (Luhmann 1996). Under present conditions, these functional differentiations have ceased to provide the political stability, as aspired to under welfare-state conditions, and economic growth, as evidenced in full-employment policies, which they once appeared to guarantee. Instead they are more likely to be regarded as obstacles to further innovation and wealth creation, regardless of wealth distribution. Even when viewed in the most favourable light, these functional differentiations of society are still perceived to be risks, if not outright obstacles, that must be carefully managed if innovation is not to be inhibited. Governments are exhorted to restrain themselves – by creating the right climate for innovation. It is argued that de-regulation is the best investment.

The rapid development of new information and communication technologies has created the technical preconditions for these far-reaching social changes. Most obviously it has provided the means by which global capitalism has been able to transcend particularities of all kinds – national, cultural, social, even individual. But it has also had more radical effects that tend to dissolve existing forms of systemic differentiation. These new technologies are themselves technically transgressive as demarcations between the mass media, voice and data transmission are eroded. The loosening of bureaucratic restrictions has led to the convergence of the telecommunications industry with the computer and entertainment industry. But, in a more fundamental way, these technologies have helped to undermine national and institutional boundaries; they have undermined established social hierarchies, moulding these hierarchies into lean organizational shapes and flat, geographically dispersed, structures.

Inevitably, individual careers and life-courses, the meaning and place of work, including 'the corrosion of character' (Sennett 1998), and the occupational structure have been profoundly affected. Public space has been criss-crossed by new networks seemingly 'privatized' and 'individualized' and also by virtual communities. The instrumental has acquired an expressive dimension and the expression of emotions, of intimacy and of hitherto private relationships has become instrumentalized. Who is driving what is not only difficult to ascertain under such conditions, but becomes a meaningless question.

But neither information and communication technology in particular nor science generally are at the heart of these social re-configurations – even if the pushes and pulls, loops and back-loops, are reinforced through scientific and technological developments. Social change is no more driven by scientific change than science is submissively shaped by society. This is why we prefer to use the concept of co-evolution. The observed changes in society and changes within science and technology clearly share a number of parameters that suggest similarities in the operation of the underlying forces. If co-evolutionary processes are at work in what we describe as Mode-2 knowledge production and in a Mode-2 society, they manifest themselves in clusters of characteristics in which new sets of perceptions, attitudes, outlooks, assumptions and rationalities coalesce with altered social practices and institutional constraints. None of these can be said to be prior to the other; nor do they reflect simple cause-and-effect relationships. Rather, these clusters or bundles are made up of elements which are typically linked in a self-organizing mode. Yet they share a number of parameters.

The Inherent Generation of Uncertainties

When Keynes drew attention to the key role played by uncertainty in the functioning of capitalism of his day, he sought to distinguish between the notions of uncertainty and of risk. For him, uncertainty was a state in which individual actors find it impossible to attribute a reasonably definite probability to the expected outcome of their choice (Keynes 1937). Rather than seeing in choices and in the risks they may carry the opening of a threatening sphere of possibilities, for Keynes uncertainty was inherent in economic life. Conceptually the notion of risk amounted to an attempt to curtail uncertainty, by attempting to assign probabilities to expected outcomes. Risks were the calculable and hence – within the limits set by probability theory – controllable islands in the sea of uncertainty. Or, to put it more simply still, risk was seen as less threatening than uncertainty. However, risk is now used in a different sense, to denote incalculability and hence uncontrollability. As a result 'risk' has come to be equated with 'danger'. This contemporary meaning of 'risk' has to some extent eclipsed the more fundamental importance of 'uncertainty' as an inherent feature of both knowledge production and social change. According to Luhmann, the distinction between danger and risk is based on the latter having been brought into the realm of calculability, and hence made subject to some measure of human decision-making, however risky, unpredictable and uncontrollable the outcome may be in the end (Luhmann 1996: 5–6).

Therefore the willingness to take risks remains a crucial element of human action and decision-making – although it has become more difficult to identify who is taking risks on whose behalf, to disentangle the complexities of decision-making or to attribute blame and legal culpability. Risk implies the 'putting up for disposition' of a given state or condition, out of which either gain, opened up through novel opportunities, or loss may result. There can be no doubt that the discussion of risks associated with science and technology merits the public attention it has received. But the contours of the Risk Society reflected in public controversies are shaped by demands for greater democratic participation in the future development of science and technology. Public discussion focuses on issues of greater accountability of science and technology and on appropriate measures and methodologies which would permit us to 'predict (and control) the unpredictable', namely to find ways of enhancing the positive benefits emanating from science and technology while preventing negative and harmful effects. In order to achieve this, technology and risk-assessment procedures are actively promoted. Similarly, 'social techniques' – such as consensus conferences, focus groups and involvement of opposition groups in sensitive regulatory procedures – are seen as potentially reducing public opposition to certain types of scientific and technological research.

But in highlighting the inherent generation of uncertainties as one of the characteristic co-evolutionary clusters in the functioning of knowledge production and in the working of a Mode-2 society, we do not speak about risks in this conventional sense. In a deeper sociological sense the key change is not so much the empirical fact that new kinds of risks are being created (although there are novel risks dramatically exemplified by the spread of the radioactive clouds, produced by the accident at the Chernobyl nuclear power plant, unhindered across national boundaries in Europe). Rather, at the turn of the twenty-first century both individuals and societies have come to understand themselves and to define many of their problems in terms of the technologies and semantics of risk (Ewald 1986: 147). The impossibility of defining an 'objective' notion of risk and distinguishing it from a 'subjective' one is now widely accepted among researchers into risk, and underlines a crucial shift in what is considered to be the relevant question. Historically, the emergence of the concept of risk implied that some of its unwanted and potentially harmful consequences could be contained or at least compensated for or ameliorated. We do not accept Beck's argument that our ignorance and inability to know, calculate and control risks inevitably redirects us back into a state where avoidance of risk becomes the only available strategy. If risks are indeed re-converted into dangers and submerged into uncertainties (which, by definition, cannot be foreseen, calculated or controlled) it may well be that what needs to be re-examined or even re-defined is the role of experts. In line with a new semantics of risk-sharing it may be necessary to promote socially distributed expertise, an argument which will be discussed in greater detail in chapter 11.

The inherent generation of uncertainties in both science and society is one of the crucial elements in their co-evolution. However, it is important to recognize that the new uncertainties in knowledge production are not simply a result of the intensification of the old Socratic maxim that 'to know more is to know what we don't know'; nor is their cause technological and economic globalization. Rather, on a historically unprecedented scale, both science and society have opted for the production of the New in an open-ended process of moving towards a plurality of unknown futures. 'Research' is primarily valued as the driving force behind economic competitiveness. As such its focus is the innovative potential of discovering the unknown and bringing it to fruition through a consciously designed and intentional process of (managed) innovation, even if no central authority is in charge – in contrast to 'science' which also embraces the systematization of knowledge and its transmission through teaching, and so requires the maintenance of complex (and conservative) infrastructures. Uncertainties abound in the process of 'research'. In François Jacob's words, it is 'le jeu des possibles' – in the sense that its outcome is not known, or knowable, before it has been achieved (Jacob 1981). The process of research and its multiple practices are steeped in uncertainties which reflect the exponential increase in the number of different directions which can profitably be explored by researchers.

The modern research enterprise has become a gigantic and unique kind of innovation machinery, simultaneously enhancing scientific creativity and selectively filtering which ideas, newly discovered phenomena or novel methods and techniques should be taken up and developed. It is not simply that the inexorable accumulation of scientific discoveries continuously creates new possibilities providing hopeful answers to the equally inexorable supply of human demands, wants and desires. New scientific breakthroughs, particularly in the bio-medical field but increasingly across many different research sites, open up new arenas where social choices can, and must, be made – as well as leading to painful dilemmas which otherwise could not have arisen. The right to know and, conversely, the right not to know have become burning issues for individuals who are faced with having to make novel choices about their medical treatment, perhaps in the area of preventive medicine or of 'alternative' medicine. The public outcry about the possible consequences and potential benefits of the successful cloning of 'Dolly' the sheep was a dramatic demonstration of the staggering proliferation of options – some imaginary, others embodying more realistic assessments of future developments. Frankenstein's shadow looms large.

But, if the history of previous debates about ethical dilemmas and the search for action guidelines when confronted with new scientific-technological options contains any lessons at all, they are, first, the attraction of actively pursuing what has become feasible and, second, the continual transgression of techno-ethical frontiers. This history, while illustrating the contingency of the possible and the allowed, neither supports nor undermines the research imperative 'whatever can be done, will be done.' The extent to which society actually takes up, embraces, adopts and diffuses new products, practices or techniques is still heavily dependent on a large number of selective filters as well as on the contingency of conditions which vary with the beginnings of each trajectory of innovation. 'Failed' innovations and 'successful' ones, when confronted and judged side by side, contain important lessons about the nature and degree of uncertainties and the contingencies that accompany them. They should constrain any retrospective storytelling.

Societies – like our own – which have embraced, in ideology and in practice, innovation as a new religion rooted in a continuous drive to bring forth the New have not only accepted – albeit as an inevitable side-effect – a certain measure of Schumpeterian 'creative destruction', they have also acknowledged in a deep sense the necessity of living with uncertainties. The accumulation of uncertainties affecting social choices and behaviour, individual life-styles and identities is unending. This accumulation of uncertainties reflects the equally inexorable increase in the number not only of notional options but of actual choices, although these options and choices may be constrained in novel ways. None of these uncertainties can be limited from the start, or factored out, because both science and society have chosen the relentless pursuit of the novel through innovation. The generation of uncertainties is as inherent to, and endemic in, research as it is to contemporary life. Among these uncertainties coexist the positive and the negative. In a typically reflexive mode the identification of the positive (and therefore to be welcomed) and negative (and hence to be condemned) is an uncertain and volatile process. However, the processes of generating uncertainties and developing coping mechanisms to reduce them are neither random nor entirely contingent. To adopt the language of self-organization, locally limited imbalances may exist but are prevented from spreading and becoming global through appropriately structured environments. These local imbalances, nevertheless, are perceived to be, and experienced as, uncertainties, which provoke counter-actions and coping mechanisms, which in turn begin to condition each other and reproduce within a closed and reinforcing system.

The Pervasiveness of a New Economic Rationality

The second parameter in a co-evolving process of science and society is the pervasiveness of a new type of economic rationality which acts as a principal filter in selecting, constraining and coping with the ever increasing flow of ever new uncertainties and options generated through research as well as in social life. It is an economic rationality that is not simply means-and-ends related or profit-oriented in a familiar way. Rather, this new form of rationality resembles the 'futures' instruments in financial markets whereby economic activity derived from first-order operations rooted in material production and exchange is displaced onto a second-order level where abstraction and speculation predominate. Potential values assessed, and profits and outcomes which have not yet materialized, become tradable in their own right; 'futures' objects assume a separate existence from actual outcomes. As a result, short-term investment and calculation of profit on the one hand and on the other a derivative economic logic displaced onto this second-order level operate jointly and/or disjointedly. The much analysed and occasionally deplored de-coupling of material production from economic investment in financial markets and the increase in free-floating speculative capital are both reflections of this new kind of economic rationality. When functional differentiation prevailed, each societal sub-system was also characterized by its own form of rationality – and economic rationality was essentially limited to the sphere of economics, although money constituted a generalized means of exchange. The onset of de-differentiation has led not simply to the wider dominance of this new economic rationality, because of the ease with which the boundaries of formerly discrete sub-systems can now be crossed; indeed this dominance has brought its own problematic, which was discussed earlier in this chapter. More significantly, this volatile second-order economic rationality appears to meet a general need because of the search for a generalized means of exchange into which all other kinds of exchange can be converted and re-converted.

Research is affected in several ways and begins to display its own variants of the more general mechanism described above. For example, for some time industry has considered it to be too costly routinely to undertake basic research in-house. Instead it prefers to screen the research produced in universities and only to become involved when it is judged to be profitable in a conventional sense. There has been a noticeable convergence between the paths of basic research and the potential of future technological applications. But this convergence has not taken the form of a simplistic predominance of more applied forms of research. It reflects the immanence rather than the instrumentality of modern science. Today insubstantial promises, which are based upon a potential that is difficult to assess properly and which will take time to develop fully but which are amplified through the media, excite the imagination of industry and the public and influence decisions about which parts of basic research are to be funded and which lines of inquiry are to be pursued (although, in 'objective' scientific terms, one may seem just as promising as the other). Collusions of interest emerge almost unaided which tread a thin line between authentic belief in the future potential and mere rhetoric of 'selling' a particular line of research to politicians and the public. Promises come first, not only simply in chronological sequence, but in order to instil and stimulate demand which later will underpin a market. The frequency with which researchers adopt 'sales' techniques in their attempt to obtain funding for what, in fact, are mere 'options', possible 'futures' fallouts or spin-offs of unknowable research results, is increasing. It is not mere cynicism, nor merely a rhetorical 'selling' exercise, but a genuine and pervasive feature of the new economic rationality that manifests itself within science just as it manifests itself in society.

Time: The Future as Extended Present

The third co-evolutionary parameter is the crucial role played by expectations or anticipations. Actions, choices and decisions are positioned on a temporal axis linking the present to the future. In a temporal sense, they extend the present horizon into a future which is uncertain. But the future is linked to the present by means of an imaginary time-space filled by a potential presumed to be open and responsive to human action, desires and fears. Expectations and anticipations comprise potentialities as well as risks, both of which influence choices and actions about how best to manage future uncertainties. Such expectations and anticipations, of course, have always existed but until recently the temporal distances involved were generally more long-term; a good example is the belief in an afterlife. What has changed is that the future has been dramatically foreshortened and is now experienced as an extended present (Nowotny 1994). As a result the conviction that the future can be shaped, provided it is imagined and anticipated correctly and provided the correct decisions and actions are taken in this light, has grown much more powerful. This helps to explain the rush towards 'visions' and the identification of trends and mega-trends. Superficially these 'visions' and trends may appear to be an extension of the mania for 'future studies' and for futuristic scenario-building in the 1960s and 1970s, but they are more reflexive in the sense that they deliberately reach out for potential allies in their own ongoing construction. By depicting 'how science will revolutionize the twenty-first century' readers are invited to transform themselves 'from being passive observers of Nature to being active choreographers of Nature' (Kaku 1998). It is argued that there is now a consensus among scientists that, in broad terms, the future potential contained in matter, life and the mind represent the three most exciting research areas if the aim is to move beyond 'discovery' to 'mastery'. Nevertheless, the planetary civilization of tomorrow still has to become persuaded of its future existence in the making. Imaginative anticipations must take over from foresight calculations. The old belief in (technocratic) predictability may be being eroded because of the growing recognition that the world is a non-linear place and full of surprises, but the belief in the efficacy of other forms of anticipations and in the self-fulfilling prophecy of expectations is on the increase.

The prevalent instrumental-utilitarian attitude towards science and technology intensifies the sense of the future as an extended present – but with problematical results. Science and technology once played a key role in helping to stabilize liberal Western democracies by underwriting the promises of future economic growth for all, but this has become frail and fragmented. The benefits and the increased wealth that flow from science and technology accrue to different social groups and strata in a highly uneven manner. This is hardly surprising, given the lack of social consensus about how these benefits and this wealth should be distributed or re-distributed. But deeper changes are also taking place. The increasing number of options flowing from the accumulation of uncertainties now influences at a much earlier stage in the process of research and innovation socially generated selection criteria. This is particularly evident in the bio-medical field. On the one hand some diseases are effectively untreated because the pharmaceutical industry is not interested in a search for drugs which will benefit only a small market; on the other there is a huge demand for medicine and medical care in countries of the Third World which do not possess the financial resources needed to create such a market. Even when financial considerations are put to one side, the customization of research products means that benefits accrue to individuals and groups in an increasingly differentiated and uneven manner. As a result the demand for, and availability of, life-style enhancing, or so-called comfort, drugs increases while public health authorities face difficult decisions about the rationalization, and rationing, of health care. Already sharp dilemmas about who should benefit from which segment of a wide spectrum of medical treatments and services, ranging from the inexpensive and low-technology to the costly and high-technology, are further sharpened by the exponential expansion, and elaboration, of these treatments and services. Research-induced innovation makes things worse. In this case access to the best (and generally most expensive) that science and technology can provide is regulated by the economics of health care; in other cases access is regulated by levels of educational achievement and the availability of information to individual customers/citizens. In all cases the result is a highly uneven and differentiated distribution of benefits, which will be exacerbated by the design of individualized 'biographies'.

More generally, however, the temporal dimension of a co-evolutionary process does not necessarily lead, as the term co-evolution suggests, to a synchronization of the major strands that shape this process. On the contrary, the complexity of the process induces frequent ruptures and dislocations. There are many instances of de-synchronization, both within institutional levels and between institutional and individual levels. For example, de-synchronization – or dissonance – is apparent in the contrast between, on the one hand, recent enthusiasm for research in fields like cosmology and astrophysics or in many areas of the life sciences and, on the other hand, the despair about our inability to find solutions to pressing social problems. This is not simply further evidence of the culture gap between science and technology's ability to offer solutions and society's reluctance to adopt them. Rather, it is a disturbing sign of the de-coupling of human creativity and institutional ingenuity. If the turmoil on one institutional level fails to be acknowledged and addressed on another level, or if there is an acute sense of crisis in one domain and a feeling of confidence and comfort in another, there is a danger that the processes of co-evolution may break down. Because acceleration is one of the dominant characteristics of the temporal dimension of co-evolution, it becomes even more urgent to pay attention to different speeds and coping mechanisms.

Space: The Flexibilization of Distance

The fourth parameter is the altered perceptions and functions of space and distance. Information and communication technologies have reduced – even eliminated – distance. Globalization can also be described as a phenomenon that has led to the compression of space, time and distance. The creation of virtual reality has made it possible to manipulate time and space according to the whims of the imagination and the social needs of the users. But distance, like the hold on the future through an extended present, is also becoming increasingly ambiguous and flexible, by collapsing and merging the global with the local. Globalization has vastly enlarged, and also homogenized to a significant extent, the space in which multinational corporations operate. On one level it has led to greater concentration; on another, globalization has made it possible to reduce the size of organizational units and stimulated decentralization. Today the ability to realize global potential depends on connections with and between highly decentralized and locally operating groups. Also, while distances may have been shrunk, this compression of space has also reshaped our perception of what it is to be 'local'. The local becomes magnified – to compensate for the altered sense of proportions; or it is extended and enlarged – as if to negate all distance.

Recent work in social studies of science has emphasized the importance of 'locality' in scientific practice and in how research operates. The traditional image of science as the representation of 'universality' embodied many of the dualisms inherited from the Enlightenment; now the local character of science-in-the-making has led to new questions. Through what processes are concepts, techniques and skills transferred from one local context into another, and what happens to them when they are transferred? Concepts that are transferred may take on new meanings; they become concepts that 'travel' in the sense of changing both over time and through space. But such concepts also need to be fertilized – and stabilized – by other, material and institutional as well as cognitive, conditions if they are to become productive. When we referred in *The New Production of Knowledge* to the many different sites in which knowledge was now being produced, we were emphasizing both a specific kind of 'locality' and the heterogeneous nature of these sites. But there are many processes at work, some producing greater homogeneity and others greater heterogeneity. When distance becomes flexible, it works in many different ways. The local embeddedness of research practices highlights the variation of research fields and diversity of research groups. Reliable knowledge, the totem of scientific objectivity and validity, can be broken down into many different local components. These components defy the image of a 'unified' science, and instead reveal the fascinating processes through which locality is stitched across flexible distances to other localities.

But distance is not only a geographical and cognitive concept; it is also a social one which is strongly related to trust. How social distance, while influencing scientific-technological developments, is also altered by them remains to be explored. But two, possibly suggestive, case studies can be offered. The first is based on a study of the sociology of missile testing and shows how social distance is related to uncertainty and trust. MacKenzie draws attention to a typical pattern with regard to confidence that may show in the results of technological testing which he calls the 'certainty trough' (MacKenzie 1996). He identifies three groups. The first comprised those directly involved in knowledge production. These were the technical 'experts' who were conscious of uncertainties of which 'outsiders' with limited technical understanding were unaware. A second group comprised those opposed to the institutions responsible for missile testing. Their perceptions of risk were characterized by this radical scepticism. But in the middle of the 'certainty trough' a third group was found, comprising those who were loyal to these institutions but not directly involved in knowledge production. Ignorant of the uncertainties of which the 'experts' who were closest to the testing site were aware, and untouched by the scepticism of the 'opponents' who were furthest away, their certainty was highest.

The second insight into an altered relationship of what social distance used to stand for is provided by the well-known sociological phenomenon of 'relative deprivation' (Runciman 1966). Traditionally those who are deprived do not compare their condition with that of people who are far better off on an absolute scale, but with socially adjacent peer-groups composed of neighbours, relatives or fellow workers. Nor is deprivation measured on an absolute time-scale; it is felt most acutely when there are signs of possible improvement. What could once be endured becomes unendurable when relief is in sight. So deprivation is 'relative' in both a spatial and a temporal sense. It can be argued – although empirical studies are lacking – that this pattern of relative, but close, distance is being undermined because the traditional conceptions of social distance have been eroded, most recently but also most powerfully through the mass media. 'Imagined communities' are becoming more prevalent, but their boundaries and identities are also becoming more flexible and volatile. The disadvantaged no longer compare themselves with their socially near peers serving as a reference group of significant others, but indiscriminately with those who have been constructed as media reference groups – which includes potentially all who are famous, rich or successful.

These sociological accounts of how uncertainty, trust or a sense of deprivation vary with social distance are deeply significant. In the missile-testing case study those close in 'knew', while those in the middle 'belonged' and those furthest away on the social distance scale 'neither knew nor belonged'. Trust, therefore, becomes crucial because direct firsthand knowledge is necessarily limited in an 'expert' society and because organizational affiliations and institutional loyalties are being undermined by increasingly volatile employment patterns and the processes of de-institutionalization. However – and this is most obvious in new conceptions of 'relative deprivation' – these new variations of social distance do not correspond to the old patterns of social inclusion or exclusion. Although these old categorizations are still important in the ordering of society, the new variations are more subtle, open to the imagination and shaped by shifting images of inclusion and exclusion. The increase of social distance, in contrast to the reduction of geographical distance, tends to dissolve fixed hierarchies of status, wealth or class. And their dissolution is compensated for by a media-induced 'intimacy' that is both imaginary and temporary.

The Self-Organizing Capacity of Science and Society

The fifth, and most important, parameter in the process of co-evolution is closely connected to the self-organizing capacity of science and society. What gives a system its self-organizing properties is precisely its capacity to define its own boundaries and thus constitute everything beyond itself as context. 'In the case of a social system, the interface can become an instrument of reflexivity for the system itself, thereby increasing its internal complexity' (Nowotny 1990: 230). Systems reduce environmental complexity; it follows that environment remains potentially more complex than systems. In other words, an initial boundary-condition of the system is its simplification ('systematization') of phenomena that lie outside itself. Reduction of external complexity takes place through increasing the system's internal complexity. As a result, the more complex a system becomes, the more powerful its potential for interacting with its environment.

There are affinities between this notion of self-organizing capacity and what Giddens, Beck and others have argued to be an essential component of modernization – hence the idea of reflexive modernization. Reflexivity arises in a number of different contexts. The first is as an element within the self-organizing capacity of 'local' systems operating under larger structural constraints while depending on continuous, and circular, processes of closure between imbalances and dissipation. Contrary to classical systems theory, the environment does not steer the system, precisely because the system operates under conditions of circular closure. At best the environment can disturb the system, but the system itself shapes the environment through its self-organizing capacities. In the context of the self-organization of the social world, reflexivity arises because meanings have to be constructed and attributed to social actions. As Giddens understands it, the terms developed to describe social life routinely enter and transform it – not as a mechanical process, and not necessarily in a controlled way, but by becoming part of the frames of action which individuals or groups adopt (Giddens 1992). This is closely linked to what has just been said about the impact of social distance. With the partial dissolution and the increasing flexibility of social structures, different and competing 'close' objects crowd in all the time. So the continuous construction of the 'local' becomes more important, because it provides the main reference framework in which a sense of stability and orientation can be constructed. However, it can no longer be taken for granted. This is one of the many tiny cracks in the social fabric through which reflexivity enters. Because the range of real or imagined choices has multiplied, the pathways connecting this construction of the 'local' to other localities and the choices implied in these connections are also multiplying. In short, a process of self-organization at work can be observed – the self-organization of the social.

But there is a second sense in which reflexivity enters the social world. While reflexivity exposes previously implicit, even invisible, relationships and interdependencies, social control has gradually moved from being expressed in an extrovert and explicit manner to more subtle forms of self-discipline and internalization of external forms. To a significant extent this remains true. Instead of being exerted through an external agency, social control has been internalized – either by acting on and through the self, or, in the case of organizations, by establishing self-monitoring and auditing systems that combine 'internal' self-discipline and 'external' social control. In *The Audit Society* Michael Power has questioned conventional accounts of audit, as a self-sustaining system of practical knowledge, and argues instead that the audit explosion reflects a distinctive response to the need to process risk. Audit emerges as 'a paradoxical and complex combination of surveillance and trust' (Power 1997: 134–5). It elicits a self-organizing description of the organization in terms of constant activation, as though every component of the organization were in a state of perpetual self-awareness, animation and explicitness. In her criticism of this 'terror of transparency', Marilyn Strathern remarks: 'Audit that so loves transparency conceals one thing: its reduction of complexity external to it is concealed under the banner of enhancement. It purportedly adds and augments organisations' understanding of themselves as agents. And thus empowerment, the agent's capacity to act and the skills that it involves, becomes absorbed into the internal complexifying development of those charged with the task of making sure it is visibly there' (Strathern 2000).

The circle is closed with the emergence of a new perception of uncertainty on the one hand and, on the other, new means of dealing with risk. In his great work on the civilizing process, Norbert Elias described the shift from external control of behaviour to its internalization culminating in notions of self-discipline and self-control (Elias 1937/1982). The acceptance of an audit and accountability culture by many social institutions at the end of the twentieth century can be seen not only as a response to democratization and greater accountability to consumers and voters, but also as a kind of 'civilizing effect' – or a demonstration of, in the words of Mary Douglas's title, 'how institutions think' (Douglas 1986). Social control, exerted through traditional hierarchies, works less and less well. In the absence of the active participation of those who are to be audited, and without an internalized institutional self-discipline, social control is ineffective. It can even be argued that, in the shift towards an audit and accountability culture (which can be regarded as forms of institutional reflexivity), an element of authenticity enters. The self, or the organization, is expected to conspire in its own surveillance. Social control is internalized and so transformed into self-control. At the same time it also becomes possible to shift from process to outcome. On the one hand the self is freer to define how specified objectives should be achieved; on the other hand the specification of performance is tightened. In a de-regulated and de-centralized world, the self becomes his or her own entrepreneur, free to choose means of how to accomplish goals, but less free to define the goals themselves.

In the case of science it has been possible to construct a relatively autonomous social space from which direct social control by 'external' political and religious authorities was excluded. At the same time 'internal' institutional quality control mechanisms allowed only peers to judge each other's work. The result was the growth of an academic ethos. This ethos established norms of professional behaviour and also shaped motivations, attitudes and, even, an 'inner calling' which in Max Weber's words marked the distinction of *Beruf* from *Berufung*. However, the co-evolutionary processes now at work in society and in science have abolished these distinctions, which can only be maintained in the environment of segregated spheres, not in a context of integration. *Berufung* is now no longer the attribute of a small number of elite groups, whether priests or scientists. Today everyone can hear and respond to Weber's 'inner calling' through the many different routes of self-realization that contemporary societies offer. Computer 'hacks' and technology 'freaks' are examples of such self-realizing life-styles. At the same time, as a result of the internalization of control mechanisms and the shift from direct surveillance to inner motivation, in order to become 'meaningful' and 'fulfilling' *Beruf* has taken on some of the characteristics of *Berufung*. This shift has been endorsed by management consultants; in business schools it is regarded as the key to success in modern management. Nowhere has the mingling of *Beruf* and *Berufung* been more in evidence than in the science system. Here the ascendant type is the 'researcher', who must demonstrate entrepreneurial talent (and, therefore, possess *Beruf*); meanwhile the 'scientist', still clinging to the values of *Berufung*, is in eclipse. In other words, the professionalization of science has brought its practitioners closer to other professional and highly educated groups in society, while at the same time it has successfully exported some of the characteristics of the scientific ethos to these groups.

But the spread of a self-organizing capacity does not end there. The peer-review system, so deeply embedded in the functioning (and the life-world) of science, is pushing up against its own limits and, in the process, being transformed. This is not so much evident in spectacular cases of scientific fraud – which highlight the enormous pressure under which some parts of the science system operate – but in the failure of scientific elites (Mode-1 elites) to accommodate societal demands for accountability and priority setting and to accept additional criteria of judging the quality and relevance of scientific work. A set of intermediate institutions has sprung up consisting of research councils, advisory committees and similar more or less bureaucratic bodies, which seek to reconcile the upholding of standards of scientific quality with new demands that transcend them and need to be incorporated. The difficulty of setting priorities in funding in basic research highlights how the system is struggling to embrace a kind of societal reflexivity – to which there is no alternative.

Conclusion

In the first three chapters we have argued that the development of more open systems of knowledge production, which we labelled Mode 2 in *The New Production of Knowledge*, and the growth of complexity, and uncertainty, in society are linked phenomena. Just as modernity, to the extent that it represented a culture of rationality rooted in a reductionist research, and modernization – the re-engineering of society through parallel processes of scientific and technical innovation and social reform – were intimately related (or thought to be in a Western context), so belief in simplistic cause–effect relationships has declined along with the conviction that society can be predictably planned. In their place has developed a pervasive sense of volatility and ambiguity. To highlight this shift from certainty to uncertainty, and linearity to complexity, we contrasted two accounts of future social change, one optimistic and essentially technicist and the other a darker account of Risk Society.

The links between Mode-2 knowledge production and what we now call Mode-2 society can be interpreted in two contexts. The first is that it has become increasingly difficult to establish a clear demarcation and differentiation between science and society. The fundamental categories of the modern world – state, society, economy, culture (and science) – have become porous and even problematical. They no longer represent readily distinguishable domains. The second is that both science and society (to the extent they can still be told apart) are subject to the same, or similar, driving forces. We have identified five dimensions, or parameters, of these forces – the overall growth of uncertainty, the growing influence of new forms of economic rationality, the transformation of time into the 'extended present', the flexibilization of space and an increasing capacity for self-organization in both scientific and social arenas. Both interpretations – the erosion of modernity's stable categorizations and the cumulative effect of these co-evolutionary processes – tend to the same conclusion: science and society have both become transgressive. This opens up the intriguing possibility not only that science can speak to society, as it has done with such conspicuous success in the past two centuries, but that society can answer back to science.

The key to understanding the complex articulations between the social and the scientific as a co-evolutionary process is not the impact of any specific parameter but their suggestive clustering and interdependent influence. The emergence of new uncertainties has been stimulated by a growing recognition of the potential of science and technology to bring forth new ideas, concepts, methods, products and instrumentation. In other words, it is a tribute to the 'success' of science, not evidence of its 'failure', although categories such as 'success' and 'failure' have been re-problematized by the rise of uncertainty because it has underlined the provisionality of all scientific results. Problems can no longer be 'solved' once and for all or even appear to be capable of solution in this simplistic sense. Instead they form a non-linear sequence which leads to new potentialities, and so to uncertainties, into which they are embedded. Any 'solution', therefore, merely offers a temporary reprieve – which leads on inexorably to the next 'challenge'. Schumpeter recognized that this sequence was integral to the process of innovation, but the full significance of this insight is only now coming to be recognized. It can even be regarded as a more disjointed and volatile form of Popperian falsifiability.

This more intense potentiality is further stimulated by the spread of a new kind of economic rationality, which enables it to transcend hitherto differentiated and specialized societal systems and sub-systems – which, in its turn, increases the potentiality of science. Next, expectations and anticipations are stimulated by perceptions of the future-as-extended-present with the result that potentiality tends to take precedence over actuality. Synchronization occurs next to breakdowns and de-synchronization. Potentiality is then intensified still more by the blurring of distance and its flexibility and the continuous construction and re-construction of the 'local'. But 'local' knowledge-production systems must develop self-organizing capacities in order to link up with other sites of 'local' systems. If this does not happen, knowledge cannot be stabilized. It is also in this way that reflexivity enters the social world, further enhancing the potentiality of knowledge.

Just as time and space have been effectively recombined into the more capacious category of space-time, so it has become increasingly difficult to establish a clear conceptual demarcation between science and society. Both, of course, as with time and space, can be measured and analysed separately, but both are being transformed in ways that are tending to erode their difference. All societies are now knowledgeable societies; this is true not only of societies that generate, and depend upon, innovation fuelled by advanced technologies (which, of course, are many because of the global reach of such technologies) but even of societies that rely on more traditional forms of knowledge production (which, in the light of the importance of contextualization, cannot be dismissed as inferior or irrelevant). Similarly science has burst through the boundaries of professionalization and institutionalization; 'researchers' are now socially and globally distributed. Two new factors have also intensified the erosion of the demarcation, or difference, between science and society. The first is that two generations of mass higher education have increased significantly the proportion of knowledgeable social actors; there are now many more scientifically trained politicians and civil servants, industrialists and business people, who can no longer be treated as incompetent outsiders. The second is that the reductionism of professionalized science means that the pool of potential peers has been systematically diluted, so eroding the coherence of generic scientific communities which can be distinguished from broader social coalitions. In these ways science has penetrated, and been penetrated by, society. It is in this sense that it is possible to speak of co-evolution. There is no metaphysic, no hidden hand, guiding the evolution of science and society in parallel; rather co-evolution is an aspect of coalescence.

4

The Context Speaks Back

In the first three chapters we described the emergence of novel and transgressive socio-economic and cultural forms, which we labelled Mode-2 society. We argued that the changes in scientific knowledge production, which we described in *The New Production of Knowledge*, and these other socio-economic and politico-cultural transformations are both characterized by co-evolutionary processes. Mode-2 knowledge production is emerging in the context of a Mode-2 society, although their relationship is neither causal nor linear, but reflexive and interactive. In the next two chapters the argument is developed in two ways. First, in this chapter we consider what is meant by contextualization. In chapter 5 we explore the forms taken by contextualization and its impact in various knowledge-producing institutions – industrial laboratories, government research establishments and research councils – and we take a detailed and critical look at the ways in which universities, still arguably the primary producers of new knowledge, have been affected and have – or have not – responded to forces of contextualization.

In modern times, science has always 'spoken' to society; indeed science's penetration of society is close to being a defining characteristic of modernity. But society now 'speaks back' to science. This, in the simplest terms, is what is meant by contextualization. Science has always 'spoken' to society – in the sense that it has provided a continuous flow of new ways of conceptualizing the physical and, to some extent, the social world. In its historical contest with religion, a triumphant science acquired a monopoly of describing and explaining 'reality', which both resisted and also validated human wishes, fancies and follies. Because the physical world, including its chemical and biological processes, came to be regarded as the most substantial component of the 'real world', a scientific definition of reality became ever more plausible. As a result the authority, values and practices of science permeated many other dimensions of society. The everyday world shrank to what scientists had 'discovered' and were able to exploit. The idea of 'scientific progress' not only validated but also contained (and constrained?) the idea of 'social progress' – so long as an ever increasing flow of scientific and technological novelties was maintained which matched the predictions of scientists. Everyday life was transformed and new socio-economic infrastructures constructed as a result of industralization; even the most remote village was connected to international networks of transportation and communication – first by electrification and the spread of railways and later by the telephone and radio.

These grand transformations were visible to everyone – and appeared to confirm the superiority of those who made them possible, notably scientists. When John Dewey wrote that 'science is an elaboration of everyday operations' and 'the home, the school, the shop, the bedside and hospital present (scientific) problems as truly as does the laboratory', he captured the spirit of an age in which science and technology apparently spoke an everyday language that everyone could – and did – understand (quoted in Schaffer 1997: 27). Indeed the home, hospital and bedside did resemble the places that science and technology had intended. And the status of citizens as producers and consumers within national economies depended on the ability to transform the everyday world into a place well furnished and stocked by science and technology. In the late 1920s the Austrian novelist Robert Musil, an astute and ironic observer of technocracy (and himself originally trained as an engineer), asked 'in an unprejudiced way how science came to have its present day aspect. This is itself important, since after all science dominates us, not even illiterates being safe from it, because they learn to live with countless things that are born of science...' (quoted in Schaffer 1997: 27).

Of course these new conceptualizations, both their operationalization and realization, have often depended on the availability of particular investigative techniques which have been grounded in particular socio-economic and cultural environments. So reflexivity has always characterized the relationship between science and society; it is not new. Indeed explanations of why modern science arose (only) in the West, or debates about the extent to which scientific achievements of science remain co-extensive with the economic and political power of the highly industralized countries of the West (and, more recently, also of Japan), reflect a strong, underlying current of contextualization. In this sense science has never been context-free. While the internationalization of science enjoyed a boost towards the end of the nineteenth century, the globalization of science, and especially of technology, occurred only in the latter half of the twentieth century, reaching its preliminary climax in its final years (Friedman 1999). As Simon Schaffer notes,

> at its start, it was still possible to pretend that scientific activity was properly European, both in personnel and institutions. Indeed, many Europeans justified their own sense of cultural (and racial) superiority by appeal to the apparent success of their sciences and technologies. Some modern commentators continued to describe a unique scientific method, which, despite resistance and incomprehension, had spread world-wide because it was the superior manner of establishing reliable knowledge. The imperial entrenchment of European institutions and economies elsewhere in the world promoted the spread of sciences and of their claims to world-wide efficacy. Globalisation had paradoxical effects. As the number of scientists rapidly increased and the age of European science ended, so appeals to a specific ethnocentric factor on whose force the sciences depended lost plausibility. (Schaffer 1997: 30)

Science's self-proclaimed ability to transcend different national and cultural contexts and to achieve universality, therefore, must be treated with scepticism. Whenever science, despite being rooted in local particularities, was able to transcend these constraints and produce generalizable outputs, these outputs typically had to be translated back into 'local' contexts in the form of socially desirable, culturally acceptable and economically useful goods, products, services and knowledge. This ability to transform the particular into the generalizable and back to the (improved) particular, to transform the 'local' into the universal – or, at any rate, generalizable – and back to an (enhanced) 'local', not only reflected science's inherent (and unique?) qualities and its special mission; it also provided the basis of its social power and, consequently, its institutional and professional privileges. In 'speaking' to society, science (rightly) insisted on its right to speak as freely as possible. But in too many cases scientists adopted the language of and aligned themselves to the powerful and privileged, in the process implying that scientists were unlike their fellow human beings.

Today, at any rate in developed countries, society 'speaks back' to science – although in the light of the discussion of earlier chapters it may no longer be possible to regard 'science' and 'society' as discrete or unproblematical categories (Goldemberg 1998). Science is now 'listening', in part because the boundaries separating science and society are becoming more porous. Again, this is not entirely new. Indeed, the capacity of science and technology to generate innovation has always depended on creative and interactive links between science and society, many of which may have been mundane but some of which were crucial. These links then created a climate that not only stimulated further innovation in the scientific arena, but in the wider social arena also encouraged the creation of economic wealth and emphasized the need for improved health and long-term sustainability in harmony with the natural environment. However, to emphasize the significance of such links and to attempt to articulate them, whether through structured national innovation policies or in less structured ways through popular and political expectations of science and technology, already implied that the language being spoken was that of society. The social had already entered. As was pointed out in chapter 1, although it is taken for granted that science and technology are the driving forces behind increased economic competitiveness and societal change, such dominant assumptions are themselves produced by complex interaction processes. The current emphasis on the potential of science and technology for innovation is only half of the story. It is equally important to describe and understand the impact of this mutual interpenetration on science itself.

In the latter half of the nineteenth century and for much of the twentieth century the purity of science was insulated from its technical utility, by the invention of a category labelled 'applied' science. However, today nearly all science and technology policies seek to strengthen the relationship between university, industry and government on the grounds that basic science is also a common resource which must make its own 'economic contribution'. As a result basic science has been *de facto* re-configured in the context of the knowledge-based economy. The other half of the story, therefore, concerns what we mean by contextualization of science. It is common ground that in science today there are many more actors; that more forces – social, economic, political – act on science; and that there has been an explosion of expectations about science's ability to provide useful answers to an ever increasing range of societal problems. The Fifth Framework Programme of the European Union, for instance, insists on the delivery of concrete socio-economic benefits in the shortest possible time. It is explicitly defined as 'problem-driven'. Nor, as was the case in the Fourth Framework Programme, is it sufficient to devise a collaborative project in applied research. Now the intention is to identify a target-problem, such as getting wet surfaces to stick to each other, secure the advice of end-users, such as doctors and dentists, and only then bring together the disciplines necessary to produce scientific solutions, such as chemistry, molecular biology and so on. Furthermore, the problems of production must be addressed from the outset. This point was emphasized in a recent editorial in *Nature* on the 'Dangers of Euro-relevance' (*Nature* 1999). However, despite these onerous conditions, when the first call for proposals was announced on the Internet, it was downloaded within 24 hours by more than 20,000 people in the case of those programme lines of interest to the life sciences alone.

But none of this means that context-free science – free from explicit contextual interference – does not exist somewhere in the existing institutional spaces. Although academic scientists must take into account the social and economic context when they design their research projects and/or write their grant proposals, Mode-1 knowledge production lives on. So too does the belief in its hard epistemological core and in the need to sustain functional protection provided by traditional scientific norms and the academic ethos. Bruno Latour has invoked a model in which '... society was compared to the flesh of a peach, science to its hard pit. Science was surrounded by a society that remained foreign to the workings of the scientific method. Society could reject or accept the results of science; it could be inimical or friendly towards its practical consequences. But there was no direct connection between scientific results and the larger context of society, which could do no more than slow down or speed up the advancement of an autonomous science' (Latour 1998: 208–9).

Our thesis is that a Mode-2 society generates the conditions in which society is able to 'speak back' to science; and that this reverse communication is transforming science. Contextualization is invading the private world of science, penetrating to its epistemological roots as well as its everyday practices, because it influences the conditions under which 'objectivity' arises and how its reliability is assessed. The claim that Mode 2 is a new mode of knowledge production largely depends on this transformative quality of contextualization. However, to argue that society 'speaks back' to science is not to assert that science no longer 'speaks' to society. This argument must be expressed with some care. First, it may be taken to imply that science and society remain fundamentally different from each other; secondly, it conjures up images of expert 'spokesmen' representing these two domains. Both are misleading. Science and society, as has already been argued, are mutually invasive and invaded; and, instead of 'spokesmen' communicating with each other, a much more plural and democratic environment has been created in which 'experts' have proliferated, which we have labelled the *agora* (and discuss later in this book).

That new knowledge is being produced in increasingly complex contexts of contemporary society is hardly controversial, but whether that knowledge should count as science is hotly contested. Those who contest the scientific status of new knowledge argue that when these contexts are themselves elements in the mode of knowledge production they inevitably obstruct the production of 'real' science; that the externally imposed objectives and expectations bound up in the contextualization of knowledge threaten to destroy science by undermining its objectivity, which depends on disinterestedness. According to this argument, contextualization is the opposite of disinterestedness. Because disinterestedness guarantees objectivity, context-free science must be preserved. But this argument not only invalidates the bulk of contemporary research; it also leads into what we call the 'objectivity trap' from which no escape is possible without abandoning disinterestedness or compromising objectivity. To evade this trap and validate the bulk of contemporary research, it is necessary to demonstrate that contextualized knowledge is at least as objective as uncontextualized knowledge – albeit in a different sense.

What are the conditions under which contextualized knowledge would allow such a reverse transformation to take place, while not undermining either disinterestedness or reliability? The challenge is to demonstrate that contextualization – often a reflection of shifting and unstable configurations of interests and perspectives – can actually enhance scientific reliability. When modern science began to carve out its own relatively autonomous space free from direct social and political control, including the cognitive authority of the church, it comprised the esoteric activities of small groups of practitioners who were committed to exploring the natural world guided by a set of rules and practices which they called the experimental method. However, the potential application of any insights they gained and of the results they produced to solving practical problems was always a lingering presence. Many early modern 'natural philosophers' were expected to contribute to the solution of such problems, for example, in navigation or artillery; many early scientific organizations, including the Royal Society founded in England in the seventeenth century and the academies of arts and sciences established in the rest of Europe during the eighteenth century, were originally created for instrumental motives, for example, to encourage agricultural improvement. However, the applicability of their results was neither their main purpose nor their principal motive for developing experimental scientific techniques. Their goal was to improve the state of knowledge. Their mission was as much post-religious as proto-industrial.

Yet, three centuries later, applicability – or, in policy jargon, the 'socio-economic contribution' that science is expected to make – has become a dominant theme. On the one hand society remains in awe of the apparently inexhaustible capacity of science to produce novelty (and with increasing efficiency and productivity); on the other hand there are inexorable demands that this novelty must contribute more exactly, and more exactingly, to improving present and future societal arrangements, stimulating wealth creation, informing social values, enhancing life-styles and individual choices, and helping to develop more sustainable relationships with the natural environment (while halting processes leading to its degradation). Of course, it is relatively easy to describe the ongoing process of contextualization in different research fields, by pointing to shifts in research agendas and how research priorities are set, and describing how the policies of research councils and other funding agencies are articulated and directed towards certain objectives, most of which follow the fuzzy contours and reflect the vague contents of the so-called Knowledge Society (or, at least, a knowledge-based economy). But contextualization has a second and deeper meaning which relates to our conceptions of how science 'really' works: what is distinctive about science; and what, therefore, cannot be tampered with without inhibiting science's capacity to bring forth the unending stream of novel phenomena, concepts, techniques and, above all, products. It is this second, socio-epistemological meaning of contextualization which needs to be scrutinized before we can understand properly how the first form of contextualization, the political and institutional, actually shapes the knowledge agenda and how this first form affects the second.

Once, questions about the distinctiveness of science were thought to be easy to answer. Philosophers of science established so-called demarcation criteria to distinguish between science and other domains. They developed a check-list of criteria which could be applied to science's inherent rules and methods as well as its findings. These efforts to establish clear demarcations between science and non-science, mature science and less mature science, failed. They were finally demolished by Feyerabend's demonstration that 'anything goes' and also by many excellent empirical studies in the history of science which demonstrated that in practice there were no clear criteria to determine what was 'scientific' and what was 'social'. Moreover, the picture that emerged was of far greater diversity of (and even dissonance between) scientific practices and research fields in different historical and geographical settings than the dominant account of a unified and context-free science was ever prepared to acknowledge. However, this did not justify the claim that 'anything goes', or that scientific knowledge production cannot be distinguished from the production of other kinds of knowledge. Nor did it justify the assertion that science is merely a social construction, which is no more satisfactory than rival accounts which insist that science is absolutely unperturbed by its social context (Hacking 1999).

Instead these studies suggested that science was actively engaged in what has been called 'boundary work' (Gieryn 1999). The assertions that science adopted the same forms of inquiry everywhere and that it could be represented and communicated in the same way proved to be as difficult to sustain as the myths about scientific unity and universality. The notion of 'boundary work' implies not only that boundaries are not fixed and permanent but that they need to be actively maintained. Moreover, their definition, mapping and maintenance often served a social function. At times science has been portrayed as unified; at others as diverse. At times it has been represented as the working of impersonal forces; at others as an intensely personal pursuit. At times it has been presented in a domestic light as similar to everyday activities in which all could engage; at other times it has been mystified, aloof from common sense and lay understanding, the working of brilliant minds far beyond the intellectual grasp of citizens. Social contingency and professional expediency influence the choices of 'stories' about science. When costly 'Big Science' projects such as the building of accelerators, manned space flight or new weapon systems were in need of public and political support, new arguments had to be found to convince lay society. The history of the major industrial, technological and scientific exhibitions and similar public displays of scientific and technological prowess in the latter part of the nineteenth century illustrates how many different answers have been given to the apparently simple question 'What is science?' (Pestre 1999). Defining the sciences, mapping their territory in public space, making and reshaping them in the image tailored for the specific time and the occasion, are all part of 'boundary work'. And scientists, as 'boundary workers', are actively engaged in such activities as an integral part of their scientific endeavours.

The attempt to define science in terms of near-absolute demarcation criteria may have failed. Nevertheless most scientists, even when acknowledging that science has to be defined, represented and presented to the public in different and contextualized ways, are convinced that it is possible to regard science as a separate sub-system of society in which its normative values, epistemologies, methodologies and its social and scientific practices continue to be distinctive. But even here there are difficulties. First, it is not clear what distinguishes the self-images and special pleading of scientists from those of, say, lawyers or musicians. Secondly, as we have already argued, boundaries between the differentiated segments of society have become fluid and porous. As a result it has become more difficult to regard science as a distinctive sub-system of society, clearly demarcated from other sub-systems, because all systems and sub-systems are in flux and have become transgressive. Science is no exception to this rule. As Mode-2 characteristics pervade society and knowledge production becomes more distributed, it has become necessary to explore the extent to which and in what ways these processes are affecting the core of scientific knowledge production. Is there a hard epistemological core underlying scientific knowledge production, which cannot be changed without destroying what makes science work? This image of a hard core, consisting of a unique mix of practices, methods and beliefs, is so powerful because the hard core is tightly wrapped in and protected by many soft layers. There is an intense conviction that this core must be defended because its presumed existence justifies the institutional protection necessary for science to operate (in addition to the professional privileges associated with a relatively autonomous social space).

The most serious threat posed by contextualization, therefore, appears to be to the autonomy of science. Of course, this autonomy has always been relative and contingent. But it has always been argued that science needed a social space of its own, however conditional and precarious, where cognitive and intellectual interests could be pursued insulated from direct social control and blatant political pressures, and that such a space is an indispensable precondition of efficient, effective and high-quality science. A similar case can be made for preserving other social spaces where institutionalized criticism and organized scepticism can flourish. A good example is the separation of powers into legislative, executive and judiciary branches. Indeed it can be argued that judges run a much greater risk than scientists of being overwhelmed by political pressures and economic interests. But in the case of science the issue is not simply one of protection, or insulation, but of creativity, and efficiency. Do scientists have a stronger claim to autonomy, simply because of the persistence of the belief in the hard epistemological core? Similar arguments can be made for other classical Mertonian norms – for example, disinterestedness. If researchers become too closely involved in running biotechnology firms or too anxious to become entrepreneurs in pursuit of profit, will they still be able to pursue disinterested scientific knowledge as a public good? This is not a moral question. When Robert Merton originally identified the norms he regarded as essential elements within the ethos of (academic) science, he emphasized their functionality (Merton 1942). The question which preoccupied him was how, if the ethos of academic science becomes distorted or decays beyond recognition, academic science could continue to function in bringing forth new and reliable knowledge and science could be maintained as a public good.

But the historical context in which Merton developed his norms of (academic) science calls into question their universal applicability and raises doubts about whether they identify what is unique to science. He wrote his famous essay in one of the darkest periods of the century, during the rise of totalitarian and fascist regimes when the unimaginable horrors they perpetrated were yet to come. Science, of course, continued to function under such regimes; moreover it was obliged (and partly willing) to serve their military and political aims. Merton was seeking to develop arguments about the ethos of science which would demonstrate that ultimately science could only flourish in democratic societies. Hence, the emphasis on the self-regulation of science and scientists, and their voluntary acceptance of norms which they themselves regarded as binding on the collectivity of which they were a part and which would ensure that science functioned efficiently. At the same time, Merton never questioned the (largely tacit) division of labour of his day. The pursuit of science in industrial laboratories remained outside his normative framework; nor was he especially concerned with the application of academically produced science. Thus, the contributions expected from academic scientists seemed clear and straightforward; all that the democratic system had to do was to let them work in autonomy and peace.

These famous Mertonian norms enjoyed strong support in the period immediately after the end of the Second World War. They corresponded closely to the self-image of many leading scientists in the United States, who returned to their universities after the war secure in their increased public prestige and enjoying much higher levels of funding. For scientists in other countries, notably Germany, they provided a strong legitimizing claim for future non-involvement in politics. The DFG (*Deutsche Forschungsgemeinschaft*) began after the war as a kind of scientists' self-help group (*Notgemeinschaft der Deutschen Wissenschaft*) to secure funding for research the scientific relevance of which they alone would judge in line with the ideal of autonomous science. But it soon became clear that in many countries there were powerful forces eroding the ideal of an academic community which was at the heart of the Mertonian norms, not least for the simple reason that the social and institutional realities of that community also began to change. The growing links between universities and industry – which had always existed but were given greater emphasis after 1945 – meant that science which was not connected with the military could no longer be entirely open. Some rules about secrecy had to be accepted, at least during the so-called competitive phase. In the new and dynamic research field of the life sciences, notably in biotechnology, the separation of university and industrial research broke down completely. University-based scientists routinely moved into entrepreneurial roles as part of their self-understanding as researchers. In the 1980s two laws were passed in the United States designed to boost innovation by giving incentives to university professors and government researchers to team up with outside firms, or start their own companies. The Stevenson-Wydler Technology Innovation Act was designed to encourage the exploitation of technological know-how in government laboratories for commercial purposes. The Bayh-Dole Act allowed universities, non-profit research institutes and small businesses which performed research under government contracts to apply for patents in their own name and keep the profits of their work. Both schemes are now widely copied elsewhere.

Other Mertonian norms, such as universalism, were challenged both on cognitive-ideological and practical-political grounds. For example, feminists exposed the deeply rooted and widespread gender-bias within science. They argued that to accept and apply the Mertonian norms as they stood was to connive at a wide range of inequalities – not only inequalities of access, promotion and status, but also inequalities in terms of substantive science, its priorities, methodologies and even epistemologies. In May 1997 *Nature* published an article on a detailed study by two Swedish microbiologists of peer-review scores for post-doctoral fellowship applications, which showed a heavy bias against female researchers. The article was widely read; following its publication the condemnation of the discriminatory and gender-biased practices was unanimous and instantaneous. The authors scathingly took the scientific establishment to task for not living up to its own meritocratic standards. But they also advanced the remarkable argument that 'the credibility of the academic system will be undermined in the eyes of the public if it does not allow a scientific evaluation of its own evaluation system' (Wenerås and Wold 1997: 343). In other words, the problem was not simply that the scientific establishment did not adhere to its own ideals and values; a new norm should be introduced, namely the accountability of science to a wider public that needed reassurance on the proper working of its quality control system.

As a result, organized scepticism, another Mertonian ideal, was doubly undermined by the blatant deficiencies of the peer-review system, even if no better alternative could be proposed. Not only was peer review not working properly (a defect which could be remedied); it might not be the most effective means of guaranteeing scientific quality in the new age of accountability (and contextualization). In an unprecedented move, intended to prevent the erosion of the traditional ethics of scientific publication, many journals, including all those published by the US National Institutes of Health, now require each author to sign a statement accepting responsibility for the whole content of a paper bearing his or her name. This requirement suggests such responsibility can no longer be taken for granted. In effect, then, many journal editors, in the wake of alleged misconduct and the suspected rise of fraudulent data, especially in the biomedical field, now disclaim responsibility for the scientific reliability of research findings they agree to publish.

The gap between the ideal and the actual practice of science had simply become too wide. New adjustments were being demanded following the undermining of the Mertonian norms which appeared to have become obsolete in the light of the altered institutional landscape and changes in the actual practice. But it may still be that the very existence of such norms shows that science is distinctive, although in a more limited respect. Other professional communities have their self-regulating norms, and some aspire to an ethos also designed to insulate members from undue external interference and control. Even the most profit-oriented professional communities, while sanctioning some individual profit-seeking behaviour, have rules and regulations – for example, against insider-trading. The medical profession, perhaps among the oldest self-regulating bodies, has developed elaborate norms and mechanisms to protect its interests and practices. As the professionalization of science continues, the academic scientific community begins to resemble more and more these other professional groups. What may distinguish scientific communities from these other professional communities is that science is a collective enterprise. Although its claim of autonomy certainly contains an element of individual gratification and an assertion of privilege, this claim can still be justified because it facilitates a collective endeavour, the pursuit and production of new scientific knowledge. This endeavour demands a sophisticated balance between, on the one hand, trust and co-operation and, on the other, fair and occasionally fierce competition, vindicating Claude Bernard's famous words uttered in the late nineteenth century: 'l'art – c'est moi; la science – c'est nous' (Bernard 1865/1966: 77).

'Nous' includes not only scientists as a cognitive and social community, a *Denkkollektiv* in Ludwik Fleck's phrase, but also the materiality of scientific practices – the material, human and conceptual practices necessary to undertake research. Scientific practice requires elaborate and sophisticated 'experimental systems' (Rheinberger 1994). Such systems comprise instruments, organisms, hypothetical (because theoretical) entities, as well as complex laboratory and other organizational arrangements which include people, objects, spaces and money. Perhaps what is truly distinctive about science is this mixture of the materiality of scientific practice and of these more theoretical, speculative and often formalized ideas and beliefs which have shaped scientific 'reality' – which itself is constantly being tested, validated and revised. The result is a shared view of a common reality, primarily of the natural, but to some extent also of the social world. This belief in a common reality is no longer naively imagined to be a 'mere image' or 'mere representation' of an independently existing 'outer reality'. Although science has obtained a monopoly of understanding and explaining the natural world, it cannot claim to subsume everything under its control. Other belief systems and everyday or common-sense knowledge versions continue to circulate. The craving for meaning, the desire to make sense of questions and existential events for which science does not have an answer are increasing. As science has abandoned its more grandiose claims of offering a comprehensive scientific world view, it has become increasingly mono-functional in its explanatory mode. Thus, a paradox arises. On the one hand science is clearly more successful than 'common sense' in extending, revising and discovering new knowledge about this shared 'reality'; on the other hand it makes little attempt to interpret 'what it means'. It is in this sense that science has become mono-functional. The task of 'making sense', of attributing cultural or individual meanings to the latest accomplishments of science, and the challenge of integrating new knowledge and the products, practices and altered life-styles which flow from this knowledge into an already existing social and cultural world, are left to others.

This is where contextualization enters – for a second time. If it is accepted that the processes that have enabled science to arrive at such a monopoly of definitions of the natural world and of a shared 'reality' are also social and that they have been obtained, partly, as the result of an astute division of labour, why is there such resistance to admitting that the result – the commonly accepted reality as defined by science – is also open to social shaping, to cultural meanings, to integration into a life-world which science makes little attempt to explain? (Of course, even positivistic scientists accept that their results are 'socially shaped' – but negatively and unproductively.) Why is the social world still largely excluded from being part of this 'reality'? Or, worse still, why is the idea that this 'reality' is also socially shaped entertained only on condition that any social dimensions, including aesthetic sensations and human emotions, must be explained in terms of 'natural' knowledge – for example, in terms of neuro-physiological processes? Why is the place of 'social knowledge' in the production of scientific knowledge denied, or even suppressed? Perhaps the reluctance of scientists to ask these questions can be explained by the fact that such questions expose science to unwelcome external influences and remind scientists of their own (fallible) human dispositions which they have striven hard to keep under control?

Part of the answer lies in the history of the institutionalization of science and in the history of scientists' attempts to come to terms with what came to be called 'objectivity'. From the start, modern science renounced any claim, in the famous passage of the Charter of the Royal Society, 'to meddle with politicks, rhetoric, divinity...'. Of course, with the benefit of hindsight, it is possible to see that science, at any rate that branch preoccupied with producing a better understanding of the natural world, was right to exclude the confusions of the political and social world with their religious wars, intolerance and violence. In the same early modern world there was an insatiable appetite for meaning and certainty. In the eighteenth century this appetite was partly satisfied by the Enlightenment with its message that Reason could be trusted, its culture of rationality and universalism and its faith in the continuous improvement of the human condition. Only later, in the nineteenth century when science and technology had become firmly locked into the project of modernization, did they indirectly contribute to knowledge about the social world and to the actual practice of shaping it. The social sciences, which grew up in the shadow and under the benevolent patronage of the nation-state and defined themselves as the children of the Enlightenment, attempted to emulate, and imitate, the more successful natural sciences. The modernization project itself was inspired by a strong belief that the rationality of science would spread to all other realms of political and social life. The 'success' of the application of scientific and technological knowledge to the building of an industrial society and the spread of science-based innovations helped to sustain social consensus, so crucial in liberal Western democracies. However, despite their 'success', the natural sciences are still reluctant to admit a more social view into the shared 'reality' which they have so brilliantly contrived. There remains a deep-seated fear of becoming contaminated by the social world.

The development of the concept of 'objectivity', according to scholars like Lorraine Daston and Peter Galison, was characterized by the struggle of scientists to control their own fears (Daston and Galison 1992; Daston 1998). Perhaps the greatest of these fears concerned their own imagination and feelings, which they came to distrust deeply. Repeated attempts were made from the late eighteenth to the middle of the nineteenth century to distinguish between those parts of their creative insights and imagination that could be controlled by Reason (which came to be regarded as 'objective') from those darker parts of the same imagination that tended to run out of control, spill over into passions and, therefore, could only lead to error, deception and flights of fancy. Out of scientists' struggles with their own feelings, which they regarded as dangerously 'subjective,' there emerged new notions of what was later understood to be 'objectivity' – notions which were polyvalent, multi-dimensional and historically contingent. 'Mechanical objectivity', linked to a set of scientific practices, like photography and map-making, and to the use of statistical, mechanical or instrumental means, became an ideal at a time when face-to-face contact no longer was able to form the source of trust that bound together the community of practitioners. Today's scientists have to confront different, but analogous, fears – their fear of the social world, with its imputed interests and ideological distortions, of cultural influences and of their own, subtle and not-so-subtle, accommodations to political and economic pressures. Moreover, it is no longer sufficient to create and maintain trust among the members of the scientific community. As public controversies proliferate, the trust of the public in science and its experts has also to be carefully nourished. If scientists would openly acknowledge these perceived threats, it might be possible to develop another model of knowledge production, in which knowledge becomes socially more robust.

What are the implications for contextualization – and also the idea of a hard epistemological core resistant to contextualization, while at the same time protecting the pursuit of science against undue influences impinging through the social context? In historical terms it is clear that contextualization has surreptitiously crept into what was once held to be the inner core of science, while science's more outwards-oriented parts have actively and openly embraced contextualization. However, there are still good arguments for attempting to preserve science's relative autonomous space. They are perhaps arguments for placing the alleged hard core in a wider context –

alongside other practices and even other 'hard cores' which may also deserve to be preserved. To do so, of course, would be a political decision based on cultural considerations. But we have not found an irreducible essence of science that would invalidate such a decision. Rather, the actual practice of science, instead of feeling bound by inherent rules which constrain it in a teleologically inspired path of development, might be set free to explore different contexts and perhaps to evolve in different directions. The research process would cease to be seen as a largely inner-directed or outer-directed (as the case may be) process and more as a comprehensive, socially embedded process in which all the contingencies, constraints and opportunities created by contextualization would be made more explicit and therefore capable of reflexive management. It is in this sense that we talk of the contextualization of science, as an enlargement of its scope and enrichment of its potential and not as an instrumentalist alternative.

5

The Transformation of Knowledge Institutions

In the last chapter we considered how, in general terms, society now speaks back to science as powerfully as science speaks to society. In the present chapter we consider, first, the importance of the co-evolutionary trends identified in chapter 2 for this more intense (and equal) engagement between science and society and, secondly, the impact of these general considerations on the main knowledge-producing systems and institutions.

Of course, proper consideration of the inroads that contextualization has made, as well as the resistance with which it has been met, cannot be achieved simply by putting different perspectives together. In the light of the co-evolutionary processes which were described in chapter 2 the dynamics of contextualization have to be seen in co-evolutionary terms. For contextualization to work in a more explicit, intended and managed way, new kinds of demands need to be identified by society while the sources of supply are uncovered by science. And vice versa, because society supplies additional resources, in the form of researchers and investment, while new demands from science also help to shape priorities and choices made on the side of society. Seen in such a co-evolutionary perspective the hard epistemological core can remain 'hard' and resistant only so long as it sees itself at the centre ('the pit in the peach', in Latour's phrase). Once the perspective is opened up to embrace context as well as content, the core becomes surrounded by other 'cores', some hard and some soft, positioned in different relations to each other and their surroundings, interconnected by both strong and weak ties.

Nevertheless, as in any co-evolutionary process, some remarkable convergences can be observed. On the side of science it is above all its remarkable efficacy to produce novelty (given sufficient resources, personnel, the right environment and its other important 'material' inputs like concepts, techniques, experimental set-ups) that now receives the greatest emphasis. On the side of society this efficacy is now being fully recognized; there is an explicit willingness to exploit this efficacy to pursue various socio-economic goals. The linchpin of these co-evolutionary strands, the connecting node that binds them together, is 'innovation'. Innovation has acquired an urgent, even quasi-moral, stridency. Amid all the turbulence unleashed at present, it is viewed as the crucial process for propelling a country, an industry, a company, a laboratory, a research field, a university or a national science system from its present state into the future. Indeed, without innovation there may be no future. The democratization of technology, of finance and of information – changes in how we communicate, how we invest and how we learn about the world – are said to have come together into a whirlwind today strong enough to blow down all the walls of the Cold War system (and much else), while creating a new division between the Fast World and the Slow World (Friedman 1999). This unrestrained belief in innovation which generates, binds together and helps to maintain the many interfaces that have arisen between science, technology and society, the world of finance, information and politics, and the domain of everyday life, comes at a time where several conditions shaping this co-evolution are coalescing.

First, there is the information revolution, the explosion of information and communication systems which now permeate the world of business and finance and that of research alike (for example, the US President's Council of Economic Advisers reports that more than one-third of 1997 fixed investment by American business was for 'information processing and related equipment'). A technological revolution, long heralded, has suddenly arrived – and is accelerating. Second, following the end of the Cold War which had provided a sheltered and supportive haven for much of basic research, especially in American universities, military goals and objectives have been superseded by those that relate to international economic competitiveness. In the European Union, competitiveness has been explicitly adopted, together with social cohesion, as the legitimizing rationale for its science and technology policy. Third, as has already been argued, the categorizations that formerly distinguished the state, the market, culture and science are breaking down and the process of further de-differentiation is accelerating. A radically transgressive

Mode-2 society is emerging. Fourth, 'research' is now valued more highly, explicitly and implicitly, than 'science'. Its capacity to 'bring forth the unexpected' – in short, the latest stunning results and exciting new findings of 'research' – is most highly valued by policy-makers, researchers, the media and even by the general public. Of course, 'research' cannot exist or flourish without 'science' – the institutional infrastructure, the transmission of knowledge and training of the next generation, the systematization of knowledge. But a distinction is now made between the two which shows a clear preference for the former and seeks ways to increase the efficacy of 'research' by shifting resources, setting priorities and maintaining a tight system of output and performance control. The increasing emphasis on diversity and pluralism tends to confirm this unstated, but powerful, preference for 'research'. Rhetorically, if not actually, 'science' aspires to unify. But there can be no 'unity of research'; it is diverse and heterogeneous by definition. Fifth, this preference for 'research' produces greater uncertainty – in the shape of the increasing number of potential directions it can follow and choices it can make, and also of increased risk-taking (although the darker sides of uncertainty, its volatility and instability, are often glossed over).

The coming together of these co-evolutionary strands and the overriding emphasis placed on innovation as the centrepiece of a new contract between science and society have opened up new spaces for individuals as active agents. Science and technology have not been exempt from the process of democratization, at least in Western liberal democracies, which has stimulated the flow of media-enhanced contestations and public controversies, and also articulated new demands for greater accountability and transparency in the science system. The information and communication technology revolution, which is being superimposed upon if not superseding industrial society, has enfranchised younger generations brought up under its electronic regime; it is also rapidly creating widely dispersed sites for accessing and working with information. The sense of Western individualism is further enhanced by the emphasis it currently receives in markets, state and culture. Competition in so many different fields and the closing-up of ranks among competitors encourage a phenomenon that economists characterize as 'winner takes all', so reinforcing the emergence of 'stars', brand names and reputation as forms of symbolic capital. But despite these trends favouring the visibility and empowerment of individuals, institutions continue to matter.

This is perhaps nowhere more in evidence than with scientific practices and research activities. Ever since modern science became institutionalized in seventeenth-century Europe it has carefully nourished and been nourished by the institutions that have been built up, transformed and maintained to serve its ends. This is most evident with universities as the oldest institutional home for knowledge production, but it is equally the case for the younger institutional siblings like corporate and governmental laboratories, research councils and a host of mediating institutions that have sprung up in more recent times – from research activities undertaken by non-governmental organizations (NGOs) in the environmental field to consulting and knowledge-transfer activities across the entire range of industry and university relations. Innovation is not a random process; it is shaped by history and institutions and nurtured (or the reverse) by a wide range of societal, cultural, political and economic arrangements. In this cross-cutting combination of investment strategies, of scientific and technological talent and skills, tax incentives, government regulation (or de-regulation), company start-ups (and failures) and the dynamics of the innovation process itself, the contextualization of scientific knowledge production thrives.

It is now generally acknowledged that the scientific knowledge input is only one – albeit an essential – factor in this bewildering configuration. But it has also been recognized, even by economists who were reluctant to accept that growth was not produced solely by increased inputs of capital and labour, that knowledge production has the strategic value of an intangible asset. The effect is to push science's 'hard core' in directions which increase the tendency for it to become de-centred. Mode-2 knowledge production occurs in many different sites and in many heterogeneous contexts of applications. The knowledge production system – industrial research laboratories, government research establishments, research councils and universities – has been caught up in this whirlwind of transformation. The rest of this chapter will consider how these various elements in the knowledge production system have been transformed. In our view the implications of this transformation will be felt most intensely inside universities, which will be discussed in greater detail in the next chapter. Of all knowledge-producing institutions, universities are unique in the sense that they both produce knowledge and train future knowledge producers. Moreover, they contain the strategic sites, or home-bases, of both 'research' and 'science'.

Industrial Research

To some extent, industry has faced dilemmas similar to (and partly aligned with) those confronting the state. It has been forced to make equivalent adjustments with regard to knowledge production. After 1945, as part of the post-war settlement in Western Europe, an increasingly collusive relationship developed between industry (part of it nationalized and therefore already 'owned' by the state) and the state in its social-democratic, Keynesian and mixed-economy manifestations. This collusion between state and industry's 'commanding heights' (a term that can usefully be extended beyond its Soviet origins to describe large corporations in the West during this period, especially manufacturing and large-scale service and utility companies) took two main forms. First, industry was often involved in the development of national research programmes. It lobbied the state to ensure a better 'fit' between these scientific programmes and national (as well as corporate) economic objectives; industrial leaders sat on committees which steered these programmes; and industrial as well as university and government laboratories contributed to them. Second, large corporations developed corporate research programmes analogous to national programmes in which clear priorities were established and, as far as possible, outputs were pre-planned. Occasionally this led to corporate programmes, and industrial laboratories, mimicking the processes of academic science. They produced basic science almost indistinguishable from that produced in university laboratories. This state-like behaviour was particularly characteristic of nationalized industries and utility companies, because they were state-owned or state-regulated, sharing a state-bureaucratic culture. But multinationals also felt the need to reproduce their own 'state-like' organization in terms of planning of research and development.

Today the environment in which industry has to operate is very different. Old habits, of course, die hard. Industry's intrusion into the shaping of national research programmes has, if anything, intensified. Industrial leaders are now more prominent (and powerful?) as members of research councils and programme steering groups. Indeed their legitimacy has been heightened: once used simply as an external sounding board or reference group, they have swiftly been re-defined as 'users' today (and not only in the case of near-market research). The Fifth Framework Programme of the European Union accords industrial research a special place and counts industrialists among its most influential advisers. They also play a prominent role in most 'technology foresight' exercises. However, in many countries there are many fewer long-term and large-scale research programmes dominated by a few big players. These have tended to be replaced, certainly in the political rhetoric, by medium- and small-sized firms. Medium-term and medium-scale research initiatives consist of a much greater diversity of participants. National planning of research has softened into foresight. Industry, therefore, is less able to influence the state's R&D policies – not because industry's involvement is no longer welcome or required, but because these polices are more tentative and less ambitious.

Not only is industrial research no longer subordinated to the state as the 'junior partner' in a state-run enterprise, it has also been affected by the restructuring of industry in the 1980s and 1990s. In most countries nationalized industries have been privatized. As a result of exposure to new competitive – and often globalized – forces, these industries can no longer 'afford' large-scale research infrastructures, nor can they maintain their previous commitment to in-house research. Moreover, they are no longer so heavily influenced by state models of doing research. Many other industries have also been de-regulated – with similar results. The organizational culture of large national and multinational corporations, or at any rate of the most successful, no longer corresponds to that of a mini-state (or, in the case of the largest multinational, an imperial state). Work-forces have been slashed; organizations have been de-layered; management has been decentralized (often breaking large companies up into quasi-independent trading units); production has been out-sourced (re-emphasizing the importance of the supply chain); and corporate functions down-sized. Industrial research departments have been seriously affected; some have simply been abolished; others have been re-engineered (often losing their quasi-academic characteristics in the process); others are expected to pay their way within 'internal markets'; and others again have been floated off as wholly independent entities.

However, it would be misleading to give the impression that the overall effect has been to produce a decline in industrial research. The opposite has happened, for a number of reasons. First, industry (albeit privatized and, to some extent, de-regulated) is now subject to powerful forms of social contextualization. Health and safety is a good example. Research in this field has taken its place alongside research into the 'core' of science and technology on which innovative and efficient production depends. It is no longer easy to distinguish between 'core' and 'peripheral' industrial processes – nor between 'core' and 'peripheral' research activities. They have become largely interdependent, not ordered by a linear chain of what needs to come first. Another, secondary, reason for this change may be that research and development in areas once regarded as peripheral to the 'core' was previously undertaken in government laboratories or laboratories established by industrial organizations and operated collectively on behalf of all the companies in an industrial sector. These laboratories have often closed, or had to take on new roles.

Second, industrial research is now much more distributed – in two ways. The first is that specialized R&D companies have emerged, especially in the small and medium-sized enterprises sector. Some of these are based on out-sourced industrial laboratories, but others are new creations, especially in areas of fast-moving science and technologies. The second is that research is now distributed back along supply chains. Often component-manufacturers and service-providers are more knowledge-intensive organizations than the big companies they supply, which may have been hollowed out into financing agencies and marketing shells. The effect of both these changes – the spreading of industrial research beyond the 'science core' into the 'social periphery', and back down the supply chain – may well have been to expand industry's total knowledge capacity.

Third, industry has been forced by the dynamics of international competition to carry the innovation process into the heartland of knowledge production itself. Firms are increasing the numbers of agreements and partnerships to which they are party. At the present time they seem to be making arrangements with a wide range of partners in the distributed knowledge production system (Dodgson 1993). This includes not only collaboration in science and technology development on which innovation obviously depends; industry is also becoming a more serious player in social science research – for example on issues related to public health, sustainability of the natural environment, risk perception, alternative dispute resolution, the operation of informal economies and the ethics of research (notably in the biological sciences). Successful innovation is now seen to require solutions to problems which demand the knowledge and skills not only of the natural sciences but also the social sciences and even the humanities. No company can these days run the risk of setting up a new factory or launching a new product line or engaging in a clinical trial without paying great attention to the social, legal and regulatory environments in which they will be operating. To deal with this, firms seek specialized knowledge. One source may be the spin-off of groups, originally developed within a company framework, which have found it more challenging and often more lucrative to offer their knowledge on a wider, freelance basis. Such spin-offs are currently pouring into the distributed knowledge production system and are available in increasing numbers to catalyse the production of contextualized knowledge.

Government Research

Parallel changes have been induced in government research establishments. Formerly, these institutions were set up to carry out research in the public interest which was thought to be not appropriate for universities on the one hand, and of insufficient interest to industry on the other. This shift in the aims and objectives of government research establishments followed similar lines to those that will be described for the research councils. They involve a clarification of their role in relation to national priorities, joint funding arrangements with appropriate industrial sectors, increased accountability and a more robust approach to the management of the research process. All these developments, it could be argued, helped to promote the production of contextualized knowledge and its advancement towards what were formerly considered to be 'core' scientific activities.

However, in some countries policy went a stage further, in the sense that formerly public research institutions were privatized and now operate in the market with other privately funded research corporations. In addition, they often compete for research funds with academic scientists through a variety of directed research programmes. Certainly in the United Kingdom, the United States and New Zealand, laboratories, formerly dependent for their funding wholly on government, have been transformed into quasi-commercial entities which are expected to make their way at least in part by offering research services to their relevant communities. It is true that, initially at least, many of these laboratories were able to retain some of their original 'customer base', and continued to supply services to a particular sector, for example the water or transport sector, or to the engineering, steel or textile industries. The transition may not have been as radical as it might appear at first. None the less, these organizations have subsequently adjusted their research profiles and appear to be surviving in more complex public-private environments. As with many research councils, here, too, formerly public institutions have been drawn into the production of contextualized knowledge. Not everyone was happy either with the notion of privatization or with the way in which it was being implemented. In some cases, groups 'bought themselves out' of public laboratories and set up their own, offering specialized services of various kinds. Consultancies spread. Thus, many former government research establishments, as well as other spin-offs from privatization, are now actors in what can only be described as a growing market for specialized – and usually heavily contextualized – knowledge. This is one area where supply and demand for knowledge are beginning to be reflexively articulated, and to the extent that new cadres of knowledge producers are able to respond to the social dimensions of their environments, these organizations contribute to the production of contextualized knowledge.

This process of contextualization has expanded scientific production far beyond the traditional knowledge-producing institutions into new arenas. The fact that many more science graduates are now produced than can be employed in academic science has had a similar effect. These new arenas now encompass corporate life more generally, including financial services, other parts of government service (apart from education, science and industry ministries), the mass media, scientific publishing, science-based investment advising and analysis, regulatory affairs and patenting, the commercialization of science and technology transfer, activist organizations, and other policy-related activities (Robbins-Roth 1998). In a Knowledge Society many institutions become 'learning organizations' – and, one might add, 'researching organizations' – in the sense that in order to survive (not only commercially, but politically) they must be able to capture, and to exploit, knowledge. To become a serious player in this game, it is not sufficient to be a consumer; one must also become a producer of knowledge. Within the widened arena of multilateral engagements, different forms of knowledge are produced, traded and, in the process, transformed. As a result, the number of players has expanded. In this sense, the mass media too must be counted as knowledge producers. They are not only actively engaged in the obvious popularization of scientific ideas and the dissemination of research findings, they are also deeply embroiled in the work of activist and community organizations, acting often as their most powerful embodiment in bringing across their concerns and issues through media representation.

To what extent such – admittedly more peripheral – knowledge-producing activities are engaged in Mode-2 knowledge production is open to debate. It has always been accepted that science and technology are diffused through modern government – whether it is to be found within the administrative capacity of ministries (not only those of science and research), in the preparation of new legislative and regulatory statutes or in the judiciary. But under Mode-2 conditions, a more radical expansion is under way (for example, through the advocacy of new policies) in the shape of research instruments that are promulgated within a much wider social arena. A new cadre of 'researchers', mainly (but not exclusively) in policy-related fields, is developing. Both the nature of their work and their status may suggest that they are 'peripheral' actors. Yet, every 'periphery' shares with its 'centre' a special kind of relationship, which may be one of diffusion or imitation but may also be one of aspiration, of moving in closer. Again, producers of knowledge become at the same time consumers and vice versa. As a result, activity at the 'centre' ultimately depends on someone in the 'periphery' acknowledging that there is a centre and taking up its activity, while the 'periphery' would cease to exist without the attraction of a 'centre'. With the 'periphery' proliferating, the 'centre' is also likely to become more distributed.

Research Councils

Traditionally, the European research council system has been run by scientists for the benefit of science. But the past two decades have seen dramatic changes in the functions, aims and objectives of many national research council systems. The strong belief in self-governance (although quite successful in the past) has come under attack. Increasingly, policy-makers in national governments feel that it fails to make the contribution expected from the science system to the solution of social problems (in Europe, notably job creation) and to the development of a strong, competitive industry in a globalizing economy. Many governments, therefore, have been prompted to orient the work of research councils more explicitly to research considered to be in the national interest. This is not the place to review all these developments, for which there is an abundance of literature. It is sufficient to note that government policies aimed at greater relevance and cost effectiveness in research emerged in a series of waves which effectively transformed the nature of research councils.

These are important developments, because research councils (or foundations) are responsible for the direct funding of a significant proportion of research. There is, however, considerable variation in terms of funding – for example, the proportion of university research funded by research councils ranges from 10 to 50 per cent. There is also great variation in the structural role research councils play in their countries, and therefore in the relative position of dependence on, or independence from, their governments. Moreover, the European Union has emerged as a new player in research funding. National research councils increasingly are influenced by the Framework Programme and need to take it explicitly into account when drawing up their national programmes. Paradoxically this effect has been very pronounced in Switzerland, which is not a member of the

EU. Because of its strong science-based industry and the importance of international markets to the Swiss economy, its researchers cannot afford to remain outside European research networks. Its research council and other bodies therefore have to take these developments explicitly into account.

The 'new utilitarian science regime', in *Nature*'s eloquent phrase, is characterized by the need to take national priorities into account, when drawing up strategies for science, developing plans to link with industry and other 'users' of scientific results (often, still in the older sense of 'industry', without drawing up policies that would also involve the 'new business') and extending the peer-review system to include experts drawn from different institutions. The principal vehicle has been the directed research programme. The willingness to introduce more problem-inspired or problem-oriented research in such core programmes (which the traditional mono-disciplinary system was reluctant to accept) has led to an increase in multi- or interdisciplinary research programmes. Some research councils, like the NWO in the Netherlands, for instance, have publicly acknowledged that interesting and exciting science can and does result from issues arising outside the strict disciplinary context (van Duinen 1998). But typically, directed research programmes have been developed – partly from the bottom up, by academic researchers, and partly from the top down, in collaboration with governments. The effect of these policy changes has been to put great pressure on research councils. Strategy papers and annual accounts published by research councils confirm this impression. Previously there was little interest in how research was actually conducted, but directed programmes now have a strong managerial component. Research is structured according to objectives, milestones, and other 'deliverables' such as dissemination and commercialization of research findings. Key words that repeatedly come up and summarize the pressures experienced are quality, relevance and efficiency.

In addition, as public funding for basic research has atrophied, new funding has become available through directed programmes, sometimes jointly funded by industry or other government departments. Government policy statements tend to convey scepticism about the contribution made by the science system to society, giving the impression that the capacity of the system is not being utilized to the full benefit of society (which is, after all, its main funder). The failure to create jobs in Europe, especially when compared to the United States, is at least partly interpreted as the failure of the science system to generate economic activity that would lead to more jobs. This notion strongly underlies government pressure on the science system in

Europe and leads to policy changes, most of which seek to interfere with the operation of research councils, universities and other elements of the system. In particular universities have come under pressure to produce more and better research – at the same time as being expected to educate increasing numbers of students (and all at no additional cost). So great has been this transformation that scientists now regularly complain that individual creativity is being undermined by the growth of collaborative, multi-disciplinary approaches to problem-solving that characterize many directed programmes. Yet the evidence is overwhelming. In most research council systems the balance of funding has shifted from the responsive mode, based on academic peer review and self-governance, to the directed mode, based on national priorities, thematic research, joint funding, managerial imperatives and a peer-review process drawn from a greater range of individuals and groups. (The undermining of the peer-review system is most obvious with the EU's Fifth Framework Programme. The programmes are designed and managed by European Commission staff. 'Experts' are only involved in the review and evaluation of these programmes, with 'experts' now being selected from a list based on self-application.) Research councils are now key institutions, linking the academic community to the research interests of government and industry.

However, although these changes in the practices of research councils appear to support our general thesis about the growing contextualization of science, that is not the end of the debate. As we will attempt to show in the next chapter, directed research programmes are instruments associated with what we have called 'weak contextualization'. Arguably such programmes were designed to solve yesterday's problems – how to plan (and, as far as possible, predetermine the outputs of) research programmes. Their aim is, first, to articulate the political context in which scientific priorities are to be established, and, second, to target resources on the most promising scientific problems and most promising scientists. But there are already signs that directed programmes are becoming less effective. They are unable to articulate the full complexity of the social contextualization characteristic of Mode-2 knowledge; they assume that reliable predictions can be made of what will constitute the most promising research; they assume, wrongly, that the link between research and innovation is direct; and, more practically, they are too unwieldy, too slow and too easy to subvert to other purposes.

Research councils, however, find themselves in a dilemma. As van Duinen puts it, the 'traditionalists' use the familiar rhetoric of purity of science, science for science's sake, and quality as the one and only criterion. The 'modernists', in contrast, accept the fact that money talks and try to reconcile traditional values with the new realities. But either way, both 'traditionalists' and 'modernists' feel endangered by the present trends and transformation. The traditional research councils are losing ground, since their arguments no longer convince when they are seen to fail in the delivery of what policy-makers aim for. However, the 'modernists' are also losing ground, because research councils are bypassed whenever direct links between universities and industry are established, driven by national priorities and European Union policies. This threat is real, and can be demonstrated by the increase in direct funding of academic research by industry, foundations and government departments (van Duinen 1998).

Two responses to this creaking of the research council system are possible. One consists in tightening controls still further, by reducing the amount of undirected research money, whether in universities' core budgets or allocated through responsive-mode competition by research councils. As a result the proportion of the research budget allocated through directed programmes and dedicated research institutes is increased. The other response, more favoured among continental European research councils, is to strike some kind of balance between bottom-up research proposals from the academic community and 'priority programmes' which are designed to have potential relevance for broader societal objectives and are often interdisciplinary. Through this process the roles of research councils are broadened to encompass both dynamics – acknowledgement of the importance of a self-governing curiosity-driven science, and acceptance of the fact that new synergies must be created with the outside world. This second response seems to us to recognize that 'strong contextualization' cannot be planned, but instead emerges out of complex communication, interaction and negotiation between a wide range of knowledge producers. The success or failure of both strategies will depend on what happens to and inside universities, the subject of the next chapter. Universities today are subject to great stress and pressure. It is in their laboratories and classrooms that these divergent forces are most pronounced; it is in the university that these competing transformations must be reconciled.

6

The Role of Universities in Knowledge Production

During the course of the twentieth century the university became the key knowledge-producing institution, at any rate in large parts of northern and western Europe and in North America. Modern society has not created any institutions to rival universities for the production of new knowledge. Other institutional foci of scientific production have been unable to match the more demanding standards of professionalization set, or the more generous resourcing enjoyed, by the university, at least since the post-war period. It has successfully absorbed or rendered obsolete such 'amateurs' as the freelance inventor, and even the autonomous intellectual is a threatened species. These other foci have been effectively incorporated in the wider university system. Once-independent research establishments have been 'captured' by it for a mixture of managerial and scientific reasons. Intimate alliances have been established between university science and the military; civil and corporate R&D have coalesced into archipelagos of like-minded knowledge-producing institutions that have come to share, to some extent, norms and values once prevalent inside the universities. While it is possible to regard these changes as unambiguous evidence of the increasing dominance the university has achieved, it can also be argued that they are, admittedly more allusive, evidence of the transformation of the university itself into a Mode-2 institution. The university then appears simultaneously as capturing, but also captured.

A generation ago, the first interpretation tended to be emphasized – the university as a triumphant, even imperialist, institution. It is still deeply ingrained in the thought and policy processes of some university leaders, civil servants and politicians, not least because the most influential interpreters themselves are often based in universities. But today, the second interpretation appears equally, if not more, plausible. In its 'take-over' of other, more democratic and more vocational, forms of higher education and its involvement in more contextualized forms of research, the university has taken on new and more diverse roles that may well be incommensurable and even incompatible with each other. The feeling of a deep crisis of purpose, administrative-managerial structure and budgeting is pervasive in many universities today. The boundaries between the university and other types of post-secondary education and other parts of the R&D system have been steadily eroded. Although this novelty has been concealed by the nominal, rhetorical and (to a lesser extent) organizational continuities of the university, its core activities and self-definition may have been deeply compromised. Before, the metaphor would have been the 'spreading' university; now it is more likely to be the 'stretched' university. Consequently, far from being inconsistent with the idea of Mode 2, the advance of the university as a place where contextualized and distributed knowledge production takes place, can itself be regarded as a component, and so confirmation, of that contextualization and its distribution.

The University as a Scientific and Social Institution

Today, universities remain the only place for the training of competent 'experts' in numbers sufficient to sustain globalized science, medicine, technology and communication systems. The traditional view of the university's engagement with science emphasized two aspects – its role as a producer of knowledge and of 'knowledgeability'. The first comes either in the form of scientific results (and other knowledge products) or in the form of scientifically trained people. 'Knowledgeability' takes the shape of both a more highly educated, and presumably more scientifically rational and literate, population and, more generally, a more enlightened society and elevated culture. In these respects there was a close correspondence between the discourses of Mode-1 science and of elite higher education. Under the regime of Mode-1 science the universities exercised scientific hegemony through their production of 'pure' research which was the foundation on which society's capacity for innovation, and the economy's ability to exploit technological advances, ultimately depended. Even if a naive linear account of knowledge production was explicitly rejected, the key role of disinterested (that is, university) science was still asserted. On the other hand, stimulating 'knowledgeability' was the key task of an elite system of higher education. The university played a leadership role through its formation of future elites, social and technical. Scientific literacy (and also, perhaps, cultural authority) would trickle down to, or be imposed upon, the general population with the help of university-educated school teachers, state officials, members of established professions and (more doubtfully) private-sector managers within a framework characterized by social and intellectual deference.

The accuracy of this retrospective view is doubtful. First, the university's scientific hegemony, never complete, is a comparatively recent phenomenon. Although evident by the end of the nineteenth century, its culmination came only after 1945, and especially with the expansion of the higher-education system after 1960, when the elite university had already come under attack. Suggestively, the scientification of the university is aligned with its massification, which may indicate unexpected strong affinities between scientific and democratic cultures. Both, after all, are aspects of the dynamics of the university, and they may be connected in two senses. The first is inherent; both fundamentally are critical, even radical, movements, sceptical of received truth and orthodox interpretations in the case of science and hostile to social exclusion and traditional hierarchy in the case of democracy. The second is contingent; only by implementing a democratic agenda, and expanding opportunities for higher education, could universities generate the additional resources they required to fulfil their scientific ambitions. There always existed a tension between the university's aim to reproduce a cultivated elite, often associated with anti-scientific (or, at any rate, anti-positivistic) notions of liberal education or *Bildung*, and its development as a scientific institution. This tension took the form both of 'culture wars' (for example, in England during the 1880s between Arnold and Huxley; in Germany with the originally higher status of the humanities which came to be challenged later by the rise of the natural sciences; or, in the 1950s again in the United Kingdom, between Leavis and Snow) and of disputes about how universities should be organized. Liberal educators emphasized the 'college'; scientific developers the department. The former stressed general education, typically at the undergraduate level, aimed at producing cultured leaders; the latter stressed specialized study by postgraduate students, matching the reductionist rhythms of science, to produce the next generation of researchers.

Any alignment, therefore, between Mode-1 science as a model of knowledge production, and the elite university as an ideal social institution, is difficult to reconcile with the historical record. There are at least as many contradictions between them as there are synergies. Certainly it would be misleading to imagine that the tensions between the social and scientific roles of the university, which are casually assumed to be sharpened by the move towards mass participation in higher education, are new. Indeed a strong argument can be constructed that the opposite is happening. The social system and the knowledge system, clearly demarcated so long as the latter was defined largely in terms of Mode-1 science, seem to overlap and lose their distinctiveness if a wider Mode-2 definition of knowledge is accepted. Just as 'society' is suffused with 'knowledge' – a phenomenon inadequately captured in the term 'Knowledge Society' which is generally used in a technicist rather than cultural sense (although other more open accounts of the 'Knowledge Society' are available such as Niko Stehr's [Stehr 1994]) – so 'knowledge' is suffused with the 'social'. If true, this has important implications for the university. Its social and scientific roles, instead of being in tension (whether between the stasis of the elite university and the dynamics of progressive science, or between the open engagement of a democratic higher education and the disengagement of 'disinterested' science), may also be starting to overlap. 'Quality', this most elusive and nevertheless essential characteristic of a university, need not be restricted to the scientific activities carried out inside the university. It can also, and should, inspire its social function, by binding the two to each other. But the social space in which such an overlap can actively be pursued, experimented with and tested, is not the administrative structure, nor budget system and certainly not anything at the rhetorical level. Rather, the place to look for it is the curriculum – with its diversity, tolerance for 'otherness' and willingness for future-oriented reforms.

However, most university policy-makers do not appear to acknowledge this possibility. Instead they see a growing contradiction between the university's role as a producer of scientific knowledge and incubator of new researchers on the one hand and its responsibility to satisfy democratic or market demands for mass participation on the other. The latter they see as attenuating the university's scientific base. Students, even the brightest and best students, are said no longer to receive a sufficiently rigorous preparation for future careers in science, because they are taught in mass institutions (or, at any rate, elite institutions struggling to compete for resources with their populist rivals); because PhDs, instead of being an apprenticeship for research careers, have become broader training programmes; and because scientific careers are becoming fragmented and, therefore, less attractive. Some of these contradictions seem real enough. In order to train a sufficient number of 'experts' to sustain a world-wide system of science, technology and research, more universities are needed, larger universities, and all of them should be competent in teaching and research at a high technical level. The message contained in this analysis is clear – more of the same, only better. At the same time, under the burden of overcrowding and understaffing, of democratization and the growing emphasis on external accountability and internal auditing, the universities lose their capacity for reform. They risk becoming huge, egalitarian and conservative with little institutionalized tolerance for 'otherness'.

In fact there is hardly any evidence to support the claim that students generally are underprepared for scientific careers. Perhaps even the contrary is the case when considering how many well-qualified graduates have to make their careers outside the academic system. The social selectivity of elite institutions in the past was no guarantee of high academic standards. According to most indicators, standards of achievement have risen despite – or because of? – the development of mass higher-education systems. While it is true that many PhD programmes now have broader goals, this reflects intrinsic factors such as the growth of new interdisciplinary taxonomies of knowledge and the 'pull' from a labour market in which the demand for narrowly focused PhDs has declined and the demand for the more generically research-skilled has grown. Other, extrinsic, factors may also 'push' from below, such as the pressures of credentialization which encourage the graduates of a mass higher-education system to undertake further courses to regain positional advantage in an overcrowded labour market, to secure enhanced status in an increasingly volatile society or even to satisfy cravings for self-realization in an anomic post-modern world. Finally, scientific careers have indeed become fragmented – as have all 'careers' defined in the mid-twentieth century – with the prospects of long-term, linear professional employment quickly fading.

However, the alleged attenuation of the university's scientific base is only half the story. Faced with many more public controversies that implicate science and technology and with activist leaders who are university graduates, it has become easy to fault the universities. It is then argued that far from producing a more scientifically literate population, the expansion of higher education has gone hand in hand with the growth of not only an irreverent but an anti-scientific spirit, bordering on a culture of irrationality. Public controversies apparently have been aggravated rather than resolved by mass participation. As a result of such accusations, the university not only has to come to terms with the erosion of its capacities and status as a knowledge producer, it is also charged with failing in its wider social responsibility to spread 'knowledgeability'. Again, there is little evidence to support such argumentation. The development of mass higher education, inevitably and rightly, has undermined habits of scientific deference. The growing proportion of university graduates in the general population has also increased the number of people who are qualified to take part in public scientific debate. And after all – why should the spirit of criticism, so dear to internal scientific debates, have no place in the public arena?

But behind such arguments and allegations that the universities have failed their mission, there are substantive changes that have turned science – and the universities as its home training ground – into a more contested domain. First, the importance of disciplinary structures and conceptual (and substantive) orthodoxies, rooted in the preferences of comparatively narrow scientific communities, have tended to become undermined by the extension of the scientific 'franchise' produced by the expansion of higher education. This can be demonstrated by the decline in social status and general societal prestige accorded to those working inside a university. Second, the growth of relativistic thinking – or, at any rate, the readier acknowledgement that there may not be only One Truth known to science alone – has undermined notions of scientific objectivity or even of reliable knowledge. Third, the very success of science (as we will argue later) has led it into arenas of more intense contextualization. This increases the likelihood of public contestation, of debates and negotiations. It may also reinforce controversy between competing 'experts' and their respective knowledge domains. The expansion of the university has certainly played an important part in transforming both the cognitive and the social contexts of science – but not necessarily to ill effect.

The Convergence of Scientific and Social Roles

The argument that the scientific and social roles of the university need not amount to a zero-sum game (the more 'scientific', the less 'social' and vice versa) is not yet widely accepted. Still less acceptable is the more radical notion that they may be mutually sustaining – and even coalescing – under Mode-2 conditions. The development of higher education and research policies in many countries has been based on the belief that it is necessary to insulate the scientific functions of the university from its social functions, often equating the former with 'elite' and the latter with 'mass' education. The intention often has been to create a clearer separation between research, in which the elite university still plays an important but no longer exclusive role, and the higher education (or, at any rate, socialization) of mass student populations where such a separation either does not exist, or to reinforce it, where it does exist, by encouraging the emergence of more differentiated systems. Of course, its effect has been as much retrospective – to maintain social-class gradations in higher-education systems – as prospective, to protect the position of science and research. Often the two agendas have simply been confused.

Three main strategies have been pursued to protect 'science' from the 'social' in the modern university and/or to preserve selective access to higher education. First, higher-education systems have been formally stratified with a small number of research-oriented universities being granted a monopoly of PhD programmes. The best-known example is the California master plan first promulgated in 1962 and revised on several occasions since, which created a three-tier system consisting of the eight-campus University of California, the California State University (CSU) campuses and the community-college sector, each with its own entry standards and range of academic programmes. Second, binary systems of higher education have been maintained in much of Europe. A clear separation has been maintained between traditional universities and higher vocational-education institutions, such as the *Fachhochschulen* in Germany and Austria and the *HBO* (higher professional) schools in the Netherlands, although there are signs of a blurring of these institutional boundaries too. In some European countries yet another form of differentiation has also been maintained – between universities and higher vocational institutions on the one hand, and, on the other, independent (or, at any rate, semi-detached) research institutions, such as the Max Planck Society, CNRS or Academy of Sciences institutes. Third, in those countries with unified higher-education systems (for example, Australia, Sweden and the United Kingdom) efforts have been made to encourage institutional diversity through selective funding policies, especially for research, and through market pressures.

None of these strategies has been without its difficulties. Political pressures have tended to dissolve hard demarcations within tiered systems. For example, the University of California's monopoly of PhDs has been breached, and substantial research programmes have developed on CSU campuses, while credit transfer systems to facilitate 'upward' student mobility have been emphasized. In Europe, even where binary systems have been maintained, higher vocational institutions are now generally embraced within the same legal frameworks as universities. Upward academic drift has not thereby been prevented. It should also be recalled that these binary systems had originally been designed not to produce a clearer separation between research-led and teaching-oriented institutions, but to cater for the further education of the separate streams of 'academic' and 'vocational' students within divided secondary-school systems. However, they have come under attack as being incompatible with democratic entitlements to higher education. The 'other' binary distinctions between universities and free-standing research institutions have tended to be eroded as the logistical inefficiencies and – occasionally – scientific sterility of the latter have come to be recognized. Finally, selective funding methodologies and market differentiation have usually failed to inhibit the development of commonality within unified higher-education systems. At their strongest they have probably done little more than slow the progress towards institutional uniformity. Some of these difficulties can be attributed to the playing out of distinctive higher-education traditions and the impact of different political environments. Generalizations are hazardous. In Germany, for example, the overt binary division is between universities and *Fachhochschulen*. Less overt is the pervasive sense of academic jealousy most universities harbour against the often nearby Max Planck institutes. While only the universities can grant doctoral degrees, working conditions for highly qualified PhD students are vastly superior at Max Planck institutes. But a covert, and perhaps more significant, binary division exists within the university between the mass-participation lecture, which reflects the imperatives of the 'social', and the professor's seminar for advanced students, which reflects the demands of 'science' or 'research'. In France too, competing principles of differentiation are at work. First, a complex pattern of *grandes écoles*, traditional university faculties and *instituts universitaires de technologie* (IUTs) has persisted, in which notions of bureaucratic elitism, radical participation (and activism) and professional formation coexist and compete. Second, this complexity is increased by the continued existence of independent research institutes, of the CNRS and other agencies, although efforts are under way to provide incentives for stronger links with universities. The result is still a largely fragmented environment, although in practice strong interconnections exist between its various institutional elements through shared personnel. In Britain institutional arrangements are apparently more straightforward. The formal binary distinction between universities and polytechnics was abandoned in 1992, and the so-called dual-support system of research (through undifferentiated core grants from the higher-education funding councils and specific programme grants from research councils) has been maintained.

But if a generalization can be justified, it is perhaps this. In the past, institutional differentiation was the product of historical variation between different types of higher education broadly defined as 'academic' or 'vocational', 'scientific' or 'professional' (the precise language has varied over time and between nations). This variation, of course, was closely aligned with social-class hierarchies. Universities typically enrolled students from more socially privileged groups than did higher vocational or teacher training institutions. Today institutional differentiation is more likely to be driven by the perceived need to protect, and enhance, research excellence. For example, a recent study commissioned by the Higher Education Funding Council for England suggested that in those countries where research and teaching are separately organized, whether this separation takes the form of independent research institutes of the Max Planck or CNRS model or of clearly delineated research universities, there is a stronger correlation between the number and impact (as measured by citation indices) of scientific publications than in countries where these conditions do not apply – for example, England (Adams et al. 1998). The conclusion drawn by some policy-makers was that research funding should be even more selectively targeted, perhaps by abandoning the dual-support system entirely.

This tension between the desire to preserve or enhance 'excellence', now defined in terms of scientific quality and research productivity (but formerly in terms of broader cultural and social-class considerations), and the need to satisfy popular pressures for increased participation, appears to confirm the existence of an inescapable contradiction between the university's scientific and social roles. Certainly policy responses have often been predicated on this assumption. As has just been argued, high-profile attempts to maintain, or promote, differentiation between research-led and access-oriented institutions have not always been successful because of the political difficulties such attempts create, although there is still room for doubt about the extent to which old-fashioned binary systems are in decline. But the segregationist campaign has been pursued on other fronts as well. In those countries, such as Britain, where traditionally universities received undifferentiated block grants from the state for both teaching and core research support, there has been a trend towards separate earmarked teaching and research budgets. This occurred partly in order to promote improved value-for-money and greater accountability, but also partly to prevent the dilution of research budgets by the demands of mass teaching (and to produce *de facto* research universities?). Furthermore, the growth of earmarked – and competitive – research funding has tended to reinforce the division of university staff into teachers or researchers. In effect, free-standing research institutes with independent budgets and separate staffs have been created within universities otherwise engaged in teaching mass student populations.

The Impact of Mode 2

Yet even these politically more discreet attempts to segregate higher-education systems into research-led universities and access-oriented institutions have met with only limited success. Generally this has been accounted for by the difficulties any selective public policies encounter in open societies; they appear to go against the democratic grain. These inhibitions may help to explain the tendency to seize on quasi-market, or actual market, solutions. Yet, in the case of the university (although the United States may be an exception), the market has also failed to produce the desired segmentation. Instead there has been a marked reluctance on the part of elite universities to concentrate on their scientific functions at the expense of their wider social responsibilities – a reluctance that can only be partly explained by either political expediency (the need to maintain, and mobilize, popular support) or the stubborn persistence of archaic notions of universality. At the same time it has proved difficult to contain research within the emergent elite sector; it has spread into other, newer and more open sectors of higher education, which again cannot be wholly explained in terms of institutional ambitions. Indeed, the containment of research (outside a few very high-cost subjects) has proved to be a failure. Not only has the number of 'researchers' within higher-education systems increased as a result of the expansion of these systems since 1960; research is now undertaken in a wider range of non-university settings which extend far beyond free-standing research institutes or dedicated R&D departments into government, business, community and the media.

To those for whom Mode-1 science is the touchstone, these difficulties and failures are merely contingent. They can only be attributed to political timidity, a refusal to acknowledge that the claims of 'science' must take precedence over the clamour of the 'social'. The solution, therefore, is to de-politicize higher education and research policies – by (re?)asserting the autonomy of the university (or, at any rate, re-establishing its privileged position within the state) and by strengthening the authority of new elites of 'experts', whether scientists themselves or state officials with a compatible orientation. Although it is not impossible to reconcile the values and practices of Mode-1 science with new patterns of industry–university relations, and some scientists inhabit both Mode-1 and Mode-2 worlds simultaneously and even comfortably, there is an underlying tension between the two which cannot be completely removed. However, in the context of Mode-2 knowledge production, these difficulties and failure of containment appear in a very different light. First, they are seen as fundamental rather than simply contingent phenomena; and, second, they are evidence of the 'success' of science and not of its decline and fall. While it was (and is) possible to contain Mode-1 science as 'objective' and 'disinterested' in the sense that it was (or supposed itself to be) de-contextualized, within a restricted number of institutions – including elite universities, – dominated by equally restricted scientific communities, it is not possible to contain Mode-2 knowledge production in the same way. Its reflexivity, eclecticism and contextualization mean that Mode-2 knowledge is inherently transgressive.

Its more specific characteristics, explained in our earlier book, also make containment impossible. First, Mode-2 knowledge production transcends disciplinary boundaries. It reaches beyond interdisciplinarity to transdisciplinarity. As a result the processes of cognitive institutionalization, which typically have taken place in and through elite universities, are weakened. Scientific communities become diffused and, consequently, the university structures of faculties and departments, institutes and centres that create and sustain these communities become less relevant. Second, Mode 2 expands the number of research, or knowledge, actors. Because of the contextualization of Mode-2 knowledge, the producers of research have become a less privileged group – and, even, a problematical category. Other actors, once dismissed as mere 'disseminators', 'brokers' or 'users' of research results, are now more actively involved in their 'production' (which itself has become a more capacious, and ambiguous, category). Third, shifting the focus from Mode-2 science to Mode-2 society, the emergence of a Knowledge Society means that a much wider range of social, economic and even cultural activities now have 'research' components. As we have argued above, many institutions are now learning-and-researching organizations – because they trade in knowledge products and because they employ many more 'knowledgeable' workers.

For the university these changes have important consequences. Under Mode-2 conditions, the distinction between research and teaching tends to break down. This happens not only because the definition of who now qualifies as a research actor must be extended far beyond the primary producers of research, but also because the reflexivity of Mode-2 knowledge production transforms relatively closed communities of scientists into open communities of 'knowledgeable' people. It is possible to argue that these 'knowledgeable' communities comprise all the graduates of mass higher-education systems, not just the minority in elite universities who have specifically trained as researchers. The new 'knowledge economy' is a *mélange* of many different kinds of activity. They include 'basic' and 'disinterested' research, but also research in its many contextualized forms. A 'knowledge economy' depends also on dissemination and even popularization; on PhD training but also on continuing education and research 'activism'. If this is accepted, it follows that the small number of universities which are research-led rather than access-oriented and which train the majority of research workers no longer occupies such a central role in this new economy. Quantitatively, they are overshadowed by mass institutions, while qualitatively the activities at which they excel, the production of Mode-1 research and researchers, are no longer necessarily the leading activities in the 'knowledge games' played in contemporary societies. Indeed, it is possible to argue that non-elite universities may be better placed to play these 'knowledge games', because they have more experience of – and less distaste for – training and building up 'knowledgeable' communities.

Certainly, the impact of Mode 2 may help to explain why elite universities are reluctant to abandon their wider social responsibilities, and why mass institutions cannot be discounted as research organizations. The scientification of society – what most people mean by the Knowledge Society – is an uncontested phenomenon. But, if the shift from Mode-1 science to Mode-2 knowledge production is accepted, it is accompanied, inevitably, by a more contentious phenomenon, the rise of contextualized science. The scientific and social functions of the university may begin to coalesce. Those who work in universities perhaps are more aware of this possibility than policy-makers, being more closely attuned to the intellectual dimensions of these changes, including curricular changes. They observe on a day-to-day basis the transgressions between teaching and research in, for example, the development of 'professional' doctorates or the growth of programmatic research (with an increasing emphasis on dissemination and reaching out to potential 'users'). On the other hand, policy-makers tend to concentrate on the logistical dimensions of the research–teaching and scientific–social dichotomies, thereby adopting increasingly selective approaches which are predicated on their incompatibility. The result may be a dissonance in strategies pursued by universities and national policies. Universities may in the future tend to adopt a more holistic perspective, predicated on the synergies between their scientific and social missions. This should translate into a greater diversity and diversification of curricula, with the aim of becoming more attuned and responsive to the combined, yet diverse, social and scientific needs and expectations of different groups of their students.

A Mode-2 University

In a Mode-2 society, the future university will need to be more of a synergistic institution – in a double sense. First, it may be necessary to delineate, and so demarcate, its activities according to anachronistic divisions between research and teaching, scientific and social roles. This entails a commitment to being an open rather than a closed, a comprehensive rather than a niche, institution. Not only does it reflect the interconnections between the university's current tasks in research and teaching, it also resonates with the long-standing belief in the unity, if not universality, of the life-of-the-mind which is part of the university's traditional life-world. Projected life-styles, reallocation of time spent in teaching and research, attempts at re-integrating otherwise fragmented parts of one's existence, present challenges that need to be met. Even if functional adaptation by itself may not be a sufficient strategy for survival and success, re-engineering may need to be accompanied by re-enchantment and the extension of the university's socio-economic, cultural and scientific roles by an opening-up of its life-world. Logistically, of course, such a transformation is a nightmare. The efficient organization and effective management of universities, as of all complex institutions, demand clear priorities – a task which is difficult to reconcile with anything-goes (and everything-going-into-everything-else). The most difficult tasks facing university leaders in the twenty-first century will be how to reconcile the university's increasingly open intellectual engagement with other policies which impinge on it, and its need to retain normative focus, strong academic and political leadership as well as managerial coherence.

Second, and much more difficult, the university will have to acquiesce in a process of de-institutionalization, because in a Mode-2 society the boundaries between 'inside' and 'outside' make no better sense than those between research and teaching. The de-institutionalization of the university presents an even more formidable challenge than finding the right kind of synergies between its different functions. The emergence of so-called corporate universities may be an example of this de-institutionalization. These take different forms. Some, like virtual universities, depend heavily on the course and qualification structures, quality assurance arrangements and academic resources of traditional universities. Typically they attempt to integrate in-company education and training with courses offered by universities – which, for example, may be granted an effective franchise to provide in-house management training programmes or may accredit existing in-house courses as components within larger academic programmes which lead to the award of degrees. This first form of corporate university is the other side of the coin to the entrepreneurial university, or the traditional university which attempts to reach out into the wider knowledge economy. Both are based on the principle of corporate–academic, and generally private–public, partnership. It is not surprising, therefore, that 'private' corporate universities and 'public' entrepreneurial universities frequently collaborate – for example, by providing joint PhD programmes.

However, corporate universities may take other forms which represent a greater challenge to traditional universities. There are a few examples of for-profit universities designed to deliver academic 'products' on a quasi-industrial basis. This differs from the behaviour of traditional universities in two respects. First, although traditional universities also engage in would-be profitable activities, these are typically at the margin, on the fringe of their core functions. Secondly, the normative context is very different. In the case of these more radically conceived corporate universities there is no pretence that autonomy and collegiality are goods in their own right, and creativity is only valued in purely instrumental terms. As a result corporate universities may be relatively unimportant sites of the production of new knowledge (although they certainly play their part in the wider *agora*, which will be discussed in chapter 13). Their articulation with the research enterprise, even in a Mode-2 environment, is weak. However, the emergence of corporate universities is another aspect of the de-institutionalization of the university.

Lurking in the background is the spectre of a potentially all-encompassing 'virtual' university, based on the exploitation of globally available knowledge products through new information and communication technologies. But this merely raises the question of how these products are generated in the first place. Institutions are still important, even if they are private corporations rather than public universities. 'Virtual' universities are essentially parasitic institutions, feeding on the work of actual universities. Their role is largely confined to re-packaging relatively traditional academic products for dissemination through new technologies. Another response to the challenge of de-institutionalization may be to revert to a more 'primitive' model of the university, based on networks of researchers, students and teachers, but elevated to and connected with the facilities offered by the latest high technologies (Darton 1999: 5–7). Although more consistent with our accounts of Mode-2 knowledge production, it still begs the question of how the sophisticated infrastructures now required to generate knowledge are to be provided. Futurism and nostalgia provide equally invalid guidance.

The university, even in mass higher-education systems, continues to fulfil two functions that depend on it being a relatively stable institution. First, it remains the most important incubator of the next generation of researchers. This is a task that no other institution is well equipped to undertake. Arguably, it is even more of a core activity than research itself, although it is difficult to envisage how researchers could be trained except in an active research environment. This function receives an even greater boost through the value attached to innovation. Among its most precious, intangible assets are well-trained and talented people. Increasingly, to assure their long-term supply becomes one of the most essential, and indispensable, contributions that universities are expected to make. But academic centres of vitality and excellence also require a certain degree of stability. The second enduring function is that of a generator of cultural norms. At the end of the twentieth century these norms may be less substantive (canons of elite ideas are incompatible with the development of democratic forms of higher education, and disciplinary orthodoxies are inconsistent with the development of a dynamic science), and more procedural – concerned, for example, with standards of intellectual conduct and, more ambitiously, the maintenance of a culture of liberal rationality (Nussbaum 1997). Perhaps the epistemological core, the source of reliable knowledge, is to be found more in these general rules of conduct than in detailed methodologies. Ideally, the university should be the promoter of both substantive, and procedural, cultural norms.

A Mode-2 university, therefore, will have to be both adaptable and resilient. It will have to be adaptable in the sense that it is able to accommodate within it apparently incommensurable but actually synergistic activities; and also to accommodate itself flexibly to new configurations of knowledge by establishing novel alliances with other 'knowledgeable' institutions. It will have to be resilient, because it must be able to provide a sufficiently stable environment to enable new researchers to be trained and cultural-scientific norms to be generated and maintained. The extent to which single institutions will be able to discharge these functions must be doubted. One solution will be to re-create higher-education and research systems – not top-down systems designed to reproduce a segmented division of labour but lateral associations to promote synergistic potential (Kleiber 1999). Another may be to create within universities new curricula and their correlates of sub-institutional forms to supplement, and maybe supersede, the more traditional research and teaching structures.

As one of us has put it, faced with many uncertain prospects and ominous signs of de-differentiation and increasing instability

> it is perhaps salutary to remind ourselves that at the end of the 18th century the university was on the brink of extinction. Stigmatised as an agent of the *ancien régime* and an intellectual backwater, it appeared unable to cope with the demands of revolutionary modernisation. The future, it seemed, belonged to radical *salons* and higher professional schools. Yet the European university was able to re-invent itself and to become the embodiment of modernity, science and democracy. So there are grounds for hope, although the scale of the challenge facing Europe's universities at the end of the twentieth century is on the scale of the challenge they faced, and overcame, two centuries ago. (Scott 1998: 455)

Even in a Mode-2 society the university occupies a central role – in two dimensions. First, in most countries universities still retain a legal monopoly of making academic awards; formal certification remains important. Second, universities remain the most important sites where knowledge gains can be consolidated, stabilized and (in some sense) institutionalized – a task which assumes even greater importance in a highly volatile society.

Conclusion

The context of contextualization has been shaped by a new – at least for Western democracies – climate in which society increasingly speaks back to science. In this process science is being transformed. Contextualization affects scientific activity not only in its forms of organization, division of labour and day-to-day practices but also deep down in its epistemological core. In retracing how messages are conveyed along these different paths, institutions and institutionally mediated activities continue to play a key role, even if the functions of institutions, and norms and behaviour within them, are being altered. In industrial R&D, governmental research and the strategic policies pursued by research councils, a wide variety of socio-economic demands has flourished and more and more cross-institutional (even transgressive) links have developed. Society speaks back by demanding innovation in a variety of ways – through national objectives, the emergence of new regulatory regimes and in the multiplication of user–producer interfaces.

The changing context of contexualization is dramatically highlighted when looking at the universities, the crucial site where contextualization or resistance to it is increasingly played out. Under conditions of the previous Mode-1 regime the social and scientific roles of the universities have often been seen in conflict. Under Mode-2 conditions we can see a possible resolution of such tension in the following sense – the so-called elite university cannot abandon the wider social responsibilities it has acquired, while so-called mass institutions cannot be discounted as research-producing institutions.

7

How Does Contextualization Happen?

In the last three chapters we have examined the context of contextualization. In chapter 4 the focus was on society 'speaking back to science'; and in chapter 5 on the transformation of knowledge production in the major knowledge-producing institutions: industrial laboratories, government research establishments, research councils and especially universities. In all these arenas the articulation of social demands and the expectations that research should yield socio-economic benefits have become pervasive. To the extent that researchers and research institutions responded to such differentiated social demand, they were moving beyond Mode-1 science into the wider terrain of Mode-2 knowledge production in the context of an emerging, and co-evolving, Mode-2 society. In chapter 6 our attention turned to the key knowledge-producing institution of the twentieth century – the university – in which so many of the tensions of this wider transformation are mirrored.

In the next three chapters the focus switches to how contextualized knowledge is produced and which forms it takes. Within the general frame of a Mode-2 society, there is strong pressure for science to move from a more 'segregated' model of interaction to one in which science becomes more 'integrated' with its social context. This is the topic addressed in this chapter. But there are at least two other criteria which mark 'how' contextualization happens. The second criterion is derived from an evolutionary perspective. It asks how demand and supply are affected by the move towards more 'integration' and whether, and if so how, the context acts as a retentive filter of selection. The third criterion is to look for the place that people occupy in knowledge production; that is, where and how people interact with others, whether they are active or passive agents, and whether they are conceptualized there as individuals or as statistical aggregates. Chapters 8 and 9 consider various forms of contextualization, from cases where – based on these three criteria – contextualization is accordingly 'weak', to cases of what we call 'strong contextualization'. But these are merely the extremes of a wide spectrum, and most other examples occupy a middle ground. There, contextualization takes place within a transaction space which is flexible, contingent and often locally dependent – what we describe in chapter 10 as contextualization of the middle range. Examples characterized by the emergence of a Mode-2 object are considered in chapter 10. We also argue that it is no longer sufficient simply to take account of the context of application in shaping knowledge production; it is also necessary to consider the more subtle context of implication.

Segregation to Integration

In her recent book on what constitute 'epistemic cultures' – the construction of machineries of knowledge construction – Knorr Cetina takes a close analytic look at the diversity in which they function. This diversity reveals, she claims, the fragmentation of contemporary science: 'a whole landscape – or market – of independent epistemic monopolies producing vastly different products' (Knorr Cetina 1999: 4). Indeed, the comparison of how two scientific fields, high energy physics and molecular biology, produce knowledge shows that they could hardly be more diverse and different from each other. She writes:

> Both are vanguard sciences at the forefront of academic respectability, both intense, successful, heavily financed – but both are also differently placed on the map of disciplines. One, experimental high energy physics, tries to understand the basic components of the universe. Commensurate with the scale of the technology used for this goal, experimental high energy physics requires a specific form of scientific inquiry, the large transnational collaboration. Besides building up an understanding of the deepest components of the universe, high energy physics also builds 'super-organisms': collectives of physicists, matched with collectives of instruments, that come as near as one can get to a – post-romantic – communitarian regime. I compare the communitarian science of physics with the individual, bodily, lab-bench science of molecular biology. The

> contrasts are many: one science (physics) transcends anthropocentric and culture-centric scales of time and space in its organization and work, the other (molecular biology) holds on to them and exploits them; one science is semiological in its preference for sign processing, the other shies away from signs and places the scientist on a par with non-verbal objects; one (again physics) is characterized by a relative loss of the empirical, the other is heavily experiential; one transforms machines into physiological beings, the other transforms organisms into machines. (Knorr Cetina 1999: 4)

Such a comparative perspective indeed brings out not only some of the features considered essential for each field, but also their essential differences. The alleged 'post-romantic communitarian' structure of high energy physics is described as a tightly managed and closed scientific and social universe. In its 'truth-finding' strategies high energy physics is comparable to the brain as a closed system in terms of information. It operates within a closed epistemic circuitry, in a world of objects separated from the environment, a world turned inward – or, better still, a world entirely reconstructed within the boundaries of a complicated multi-level technology of representations. At its core is the detector that provides the first level of these representations with all their ambiguities. Yet, particle physics is perfectly capable of deriving 'truth effects' from sign-processing operations. Knorr Cetina proceeds to describe how in the experiments the emphasis is switched from 'observing the world' (or 'care of objects') to the observing of their own components and processes ('observing the self' or 'care of the self' as understanding, observing, describing what happens inside the system). High energy physics is marked by the loss of the empirical – in experimental high energy physics, experience appears to provide no more than an occasional touchstone that hurls the system back upon itself, and 'success' may well depend on how intricately the system interacts with itself.

Paradoxically, the machines become analogous to physiological organisms. Detectors are described as moral and social individuals, who can be 'young' or 'ailing' and who display behavioural idiosyncrasies. Perhaps not surprising, the physicists socialized and working in such an environment become completely subjugated to it. Not only is the individual physicist said to be 'erased' as an epistemic subject (as is the case when authorship becomes attributed to groups of scientists that may exceed one hundred), but groups are classified and defined through their relation to a specific task and to a set of objects that forms a distinctive unit. Physicists are said to function as 'symbionts' in the experiments who draw their strength, identity, expert status and the attention they get from others from that very symbiosis.

Compare now these 'communitarian' practices, the 'management by content' and other features of this epistemic culture with the field of molecular biology. They could not be further apart. In Knorr Cetina's description, molecular biology is 'a system open to natural and quasi-natural objects' and it relies on maximizing contact with the empirical world. The body of the individual scientists working in these labs is an important sensory resource, whenever manual work demands special (bodily or sensory) skills and tacit knowledge. In contrast to high energy physics, the organisms that molecular biology works with are turned into machines – naturally occurring animals and plants have disappeared from the work-bench of biology and have been replaced by micro-organisms, cell lines and animal strains re-designed to perform as laboratory types. Organisms have thus been transformed into production sites and into molecular machines.

Instead of the 'erasure of the individual as epistemic subject', individualism is alive and well in every molecular biology laboratory. Not only are these run as individuated units, free from the necessity to rely on a central machine and its immensely complex co-ordination requirements, the individual scientist, and in particular the leader of a molecular biology laboratory, reigns supreme. The 'scientific person' *qua* individual is thoroughly sustained: work is assigned personally, in the form of 'his' or 'her' project, and when laboratory leaders describe a laboratory's work, it is done by running through current projects. Post-doctoral and senior molecular biologists may seek contributions by other scientists from inside or outside the lab to speed up their research, but in ways which preserve their status as individuals, just as authorship conventions equally uphold this status. Individual skills and the small life-world contextual arrangements in which these object-centred relationships become alive only reinforce the importance of creating a setting designed to enhance an individual's (embodied) competence towards 'their' material object. Techniques in molecular biology and related fields travel through 'packages' of arrangements that incorporate scientists and material objects and that need to be recreated in every local context. Tied as they are to a scientist's body, they nevertheless need the other, material arrangement of these local settings (which sometimes make reproduction of results difficult, when moving from one laboratory to another). Last, but not least, 'owning one's own lab' is the ultimate desire of those who wish to remain in an academic setting. It culminates in becoming a laboratory leader.

We have dwelled at such length on Knorr Cetina's study because of its direct relevance to our argument about the 'how' of contextualization. As is evident from this study, high energy physics can be said to be much more 'segregated' from society on many counts. The world of particle physics is a 'closed' world, of short-lived, transient objects which are subject to frequent and changing occurrences and decay which can only be 'seen' through the traces they leave when flying through different pieces of equipment – the detector 'sees' or 'senses' their presence and registers their passage through complicated steps involving different levels of representations. But there is not only the 'segregation' from the empirical world as we know it in other sciences. 'Segregation' is mirrored in the strong collective ties that bind the groups of high energy physicists together, necessitated and reinforced through the long-term and detailed planning efforts that precede the setting up of an experiment. It is vindicated through the convention of multiple authorship by virtually everyone who participates. The links to society, it can be argued, are few, and those that exist, as studies of the history of CERN and similar facilities show, are made entirely subservient to the goal of setting up and running an experiment (Traweek 1988; Krige and Pestre 1997).

In contrast, the main features revealed in the epistemic practices of molecular biology and its organizational settings are much closer to what we call 'integration'. The laboratories are geographically dispersed and function as individuated units. The many analogies to other 'production sites' in contemporary society are striking. So is the role of the individual scientist in the laboratory, displaying not only different bodily skills and experience but also pushing and being pushed into highly individualized career strategies. All these characteristics pertain intimately to what Knorr Cetina calls 'epistemic cultures' or practices. She does not investigate strategies of how funding is secured or how links to industry are maintained. These and other issues that we have been dealing with remain outside her epistemic focus. It is therefore all the more striking that even in such an internal view of the 'construction machineries' of scientific knowledge, their deep and unintended embroilment with society is all the more visible. For those who are ready to listen, society speaks loudly through molecular biology, while it can hardly be heard as a whisper in the case of high energy physics.

It is necessary to place the contextualization of knowledge production into a broader frame of understanding. In trying to see how cultural values in a more general sense constitute important resources that not only affect economic performance and the outcome of interaction among individuals but also bear upon scientific and technical knowledge, we want to introduce a model of 'segregation' and 'integration' of social forms of co-operation and competition which has been developed by economic historian Avner Greif. To demonstrate the importance of what he calls behavioural beliefs, he bases his model in both theoretical and historical studies. Behavioural beliefs concern actions that will be taken by individuals in situations which will never occur. They are expectations that members of a society have with respect to actions that, in game-theoretical language, are taken off the path of play. The theoretical study demonstrates how behavioural beliefs coalesce with divergent trajectories of social systems and economic organization. The model is constructed in such a way that it permits the comparison of results with detailed historical evidence available from Muslim and Latin societies of the late medieval period. It shows that diverse cultural beliefs have different implications with respect to the social structure of a society and demonstrates that whether two economies, identical in all but their behavioural beliefs, become integrated or segregated is a function of the behavioural beliefs which affect their response to a given change in the rules of the game.

Based on detailed historical studies of social organization as it developed among communities of Maghribi, largely Jewish, and Genoese merchants in the late medieval period, Greif shows how both Maghribi and Genoese traders faced a particular change in the nature of their trade. The scope of their trade expanded and they found themselves capable of becoming active in long-distance trade on a scale unimagined before. The theoretical model predicts that the Maghribi traders of the eleventh century, who held largely 'collectivist' behavioural beliefs before or throughout the process of expansion, would as a result display 'segregation' of agency relations. The Genoese traders of the twelfth century, holding 'individualistic' behavioural beliefs before or throughout the expansion in trade, would organize themselves in terms of the model of 'integration' of their relations. Briefly put, the Maghribi Jewish traders, in spite of the opportunity and economic incentive to co-operate with trade centres outside the Muslim world, neither emigrated nor established agency relations with Jewish traders from the Christian world on a substantial scale. The Genoese traders, on the other hand, who indeed held individualistic behavioural beliefs, responded in a different manner to the possibilities of establishing agency relations with non-Genoese merchants abroad. They employed them on a permanent basis, making a series of substantial organizational innovations – by inventing new forms of business associations – on the way (Greif 1997).

Historians and social scientists have often remarked on the rise of individualism in Western societies and the relative eclipse of societies organized around collectivist beliefs and behaviour. Such societies tended to be 'segregated' in the sense that while there is a high level of co-operation among the members of the group and a highly homogeneous system of values and norms that binds members together, co-operation with other social groups or individuals tends to be formalized and highly circumscribed. The relations between individual members of different groups, therefore, are characterized by a relative lack of co-operation and a low degree of competitiveness. Individualistic societies, in contrast, are said to be more 'integrated' in the sense that there is a higher level of co-operation among members of different groups and competition is more open. The two types of society also differ with regard to enforcement rules and other organizational arrangements which in the long term may lead either to path-dependent organizational innovations or to what have been called locked-in situations. Greif emphasizes, however, that his model does not seek to predict which type of social arrangement, 'segregated' or 'integrated', is more 'efficient', since such a question depends on the parameters of the model. However, it can be said that collectivist societies (and the collectivist belief-systems on which they rest) are likely to generate a more horizontal social structure. This means that within a group there is more of a symmetry of norms, roles and status, producing greater equality among its members. Because they enjoy similar status, they are expected to behave towards each other on a basis of mutuality and reciprocity. In contrast, individualistic societies tend to generate more vertical structures, in which there is a marked asymmetry of norms, roles and status and, subsequently, a greater degree of inequality. Members of social groups enjoy different status. Distinctive norms and roles are associated with these status differences – or different 'identities'.

This contrast between 'integration' and 'segregation' can be linked to the emergence of a Mode-2 society and, consequently, it may help to explain changes in the complex articulations between science and society. Changes in the belief system, in norms and values as well as in the behaviour underlying the social relations of knowledge production are a precondition for the production of contextualized science. One important change is the erosion of the collectivist belief-systems that characterize the science system and generate the norms which bind it together. The result is less 'segregation' from, and more 'integration' with, society. As we have repeatedly emphasized, the boundaries become more permeable. Under the new rules of engagement scientists are no longer expected to interact almost exclusively with other scientists; nor does wider interaction with members of other social groups occur only along predetermined and prescribed lines. Scientists now share their once exclusive systems for communicating information with these 'outsiders'. One way of putting it is to say that the rising tide of individualism in society has now reached collectivist scientific communities. Although science is still a collective enterprise and the collectivist, consensus-dependent beliefs about the essence (and the constitution) of science are still influential, individual scientists are now much freer to co-operate with individuals from other groups and to transgress established institutional and group boundaries. Information-sharing and transmission may still differ in substance (scientific content and communication styles). But expectations and actual behaviour now take place in much wider arenas or transaction spaces. Even the scientific content is opening up to research questions that depend no longer on the image of Nature whispering them into scientists' ears. Individuals from other social groups, whether members of other scientific communities, industrial partners or lay people, are now actively sought, valued and welcomed in the new game of knowledge production characterized by much more individualistic rules of engagement and potential for developing novel relationships with those who are in the rising category of the ubiquitous 'user'.

According to this 'integration/segregation' model new co-operative relationships are emerging in response to the opportunities which Mode-2 society and its economic incentives offer to science on a historically much expanded scale. New relationships (and types of social organization) are partially replacing older collectivist relationships where scientists deal mainly with other scientists (and, indeed, predominantly with members of their own discipline). The two examples presented in the beginning of this chapter – high energy physics and molecular biology, as described by Knorr Cetina – are a forceful illustration of how differently two vanguard communities may respond, but care should be taken not to re-create under a new label the barrenness of a series of dichotomies which have been disproved so often. We should listen attentively to authors like Peter Galison, who reminds us that

> Physics is a complicated patchwork of highly structured pieces: instrument makers thoroughly versed in the manipulation of gases, liquids, and circuitry; theorists concerned with the coherence, self-consistency, and calculability of the behaviour of matter in their representation of matter most finely divided; and experimenters drawing together instruments into combinations in pursuit of novel effects, more precisely

> measured quantities, or even null results. But between and among these subcultures of physics lie substantial border territories, and it is only by exploring the dynamics of those border regions that we can see how the whole fits together. Between theoretical nuclear physics and the art of the cloud chamber construction is the art of interpretation, borrowing bits of theoretical ideas, pieces of craft knowledge about films and optics, and portions of experimental knowledge. (Galison 1997: 584)

In this sense, then, physics can be said to be multi-cultured – despite the valiant attempts of both positivists and anti-positivists, who sought and found a single narrative line that would sustain the whole, not only of physics, but of science *tout court*. Yet, the seeming disorder of the scientific community, the laminated, finite, partially independent strata supporting one another; and its seeming disunification, the intercalation of different patterns of argument, are at the same time responsible for its strength and coherence (Galison 1997: 844).

The Changing Rules of Engagement

Thus, the new, more individualistic relationships are pushing towards greater 'integration'. At the same time a more vertical structure of social relationships is developing. The status, or the professional identity, of the researcher can now become linked to new and different norms, outlooks and practices, which are no longer exclusive to scientists but can also be found among non-scientists. The rise of individualism, however, makes it unlikely that only one kind of vertical structure will develop. Rather, this process of stratification will be plural. There are already now many recognized different ways of being a scientist or a researcher, and many different professional identities and ways of becoming positioned in the science system have become possible. As a result 'scientific' elites proliferate and the diversification of the science system proceeds apace. However, a number of preconditions will have to be satisfied before the shift towards 'integration' which we have identified can take place on a wider scale. The general loosening of social relationships and institutional structures is accompanied, and perhaps promoted, by the social distribution of knowledge production. As we remarked in *The New Production of Knowledge*, the social basis for this phenomenon lies in the widespread expansion of higher-education systems, and in the parallel expansion in the number of sites where recognizably competent research can be carried out. The shift towards 'integration' is greatly aided by the pervasive role of information and communication technologies. The general weakening of boundaries, blurring of identities and broadening of what it means to be involved in knowledge production implies also a general expansion on both the supply and demand sides, cutting right across what used to be a fairly distinct line separating science from society.

On the supply side the situation is dynamic. Since the end of the Second World War the success of science has lured more and more young people into the enterprise; this has happened, in part, because of the expansion of national systems of higher education. With the massification of higher education, more and more individuals have been trained to examine problems and issues in specific ways. They have not only been inducted into the general ethos of science, they have also learned the particular theories, methods and techniques of some discipline. Because the output of individuals so trained could not be wholly absorbed even by an expanding higher-education system, this expertise is now widely diffused throughout society. It has itself become a resource, available to apply its knowledge and skills to a wide range of problems. On the supply side, therefore, only a minority of graduates now work within the disciplinary structure of science in which they were trained. Most graduates conduct recognizably competent research in a wide range of organizations and environments. Furthermore, even among those who continue to work within the disciplinary structure there are many who now use their expertise where new opportunities arise. Within a distributed knowledge production system there is a much larger range of problem opportunities to which disciplinary experts can apply their knowledge. Under contemporary conditions it is more accurate to describe disciplinary science as an integral part of a much larger system of knowledge production, rather than the other way round.

Demand is equally dynamic. The success of science has also encouraged the widespread belief that scientific knowledge is a reliable source of solutions for all kinds of problems. As a result there has been a great expansion in demand for the type of specialized knowledge produced by researchers using the methods of empirical science, whether physical or social. It is hardly surprising that advanced industrial societies should develop along these 'scientific' lines, given the success of science in providing pathbreaking solutions to so many problems – in energy production and manufacturing, in the treatment of disease and the conduct of war (with its undeniably rich fallout of civilian uses). Because of its success, science has come under more and more pressure from society to deliver effective solutions to a wider range of increasingly complex problems. Through this process science is being drawn into the production of more and more contextualized knowledge – in other words, attempts to solve problems that have their origins in the concerns of particular individuals, groups or organizations, or even society as a whole.

These changes in both the supply of knowledge producers and demands for knowledge production make the shift towards an 'integration' model much more plausible. They mean that there are more opportunities to collaborate outside the disciplinary structure and to work with a wider array of expertise, including, now, that in the social sciences and the humanities, just as the many new trading opportunities which arose on both shores of the Mediterranean in the late medieval period prompted innovative forms of social organization as well as a spate of new financial, organizational and technical innovations underpinning the expansion of trade on a larger geographical scale. Of course, today there are still numerous opportunities for multi-disciplinary research within disciplinary, or Mode-1, science. Such hybrid developments will continue to occur. They are part of the inner dynamics of the science system. However, there is evidence that, while in the past multi-disciplinary, or hybridization, typically occurred between activities which were close to one another on the disciplinary map, this is no longer the case. Once distant disciplines are now combining (Georghiou 1998: 1–3). To some extent this wider hybridization across disciplines seems to prompt, and be prompted by, transfers of technically sophisticated instrumentation from one research field to another. One example is the pervasive use of simulation techniques and modelling with the help of visual image technology and techniques. As a result simulation is now considered alongside theory and experimentation as one of the three pillars on which the scientific method rests.

But a shift from 'segregation' to 'integration' requires more than altered conditions of supply and demand, as they typically arise with new opportunities. Such a shift must also be grounded in altering the belief systems upon which all social arrangements ultimately rest. This is where reflexivity enters in our account of the ongoing process of contextualization of knowledge. Much of the scholarly writing on modernity emphasizes the importance of reflexivity in social relations in producing a general loosening-up of institutions. Reflexivity draws attention to the fact that discourse becomes constitutive of the social reality it portrays. In other words, when there is a new terminology, new concepts and ideas to grasp certain features of an ever changing reality, these terms seep into social life and help to re-order it (Giddens 1992). Reflexivity becomes institutional reflexivity because it is a constitutive element of social activity in modern settings and therefore spreads rapidly and widely. It is reflexive because descriptions enter social life and transform it, not necessarily in a controlled way and certainly not as a mechanical process, but as part of a more general framework of beliefs and expectations which individuals and groups adopt. To return to the supply of and demand for specialized knowledge, it can be argued that supply and demand are reflexively articulating one another in what has become a market-place of knowledge – emphasizing, even if the term 'market' is used in a metaphorical sense, the individualized and individualizing beliefs, values and norms upon which the functioning of any market is premised.

But the past success and continuing expansion of science have transformed the environment in which demand was previously articulated. Since the end of the Cold War few governments have been prepared to increase science budgets. The collapse of Communism and the end of the superpower arms race inevitably meant that previous levels of military R&D could no longer be sustained. Instead the popularity of market-oriented research has redirected national priorities in science and technology. There has been a marked shift of attitudes towards science on the part of governments as well as in the general public mood; science is now predominantly regarded, and therefore valued, as the engine in the global race for economic competitiveness. But to manage science and technology policies in order to maximize economic growth is a much more complex task than implementing the traditional security-oriented policies of the immediate past. Under these new conditions demand no longer comes from, or is channelled through, a few well-known sources, and the state has ceased to be the primary client and consumer of R&D activities. Increasingly demand also comes from individuals, groups or organizations that have the will and the capacity to generate resources to pursue problems in which they are interested. The more complex articulation of supply and demand has stimulated the production of contextualized knowledge. Innovation, which nearly all science and technology policies aspire to encourage, is a more demanding process than research, including the most promising research, in the sense that a much wider range of factors must coalesce for successful innovation to occur. Innovation demands novel organizational forms, radical approaches to design and marketing and many other extra conditions which mainly become visible when they are absent. Neither the state nor the market by themselves can hope to supply all these disparate elements and satisfy all these conditions. But the overall direction of these developments suggests that the science (or, rather innovation) system must move towards, in Greif's terms, a model of 'integration', abandoning past habits of segregation and collectivist belief-systems of scientists.

How does the process of reflexive articulation work in the sphere of knowledge production? How, concretely, does the desire to solve a particular problem (demand) find the appropriate elements of specialized expertise (supply)? Or, conversely, how do 'experts' become aware of problems to which they might contribute? How do problem-solvers interact with their environments? This process of reflexive articulation will exhibit a complex dialectic, one in which the rules of the game, however subtle or explicit they may be, are bound to change as well.

Of course, historical or cultural influences have always shaped science. But direct, or even indirect, links between the science that scientists chose to pursue and social, economic or cultural needs were regarded for a long time with suspicion, because such links were thought to limit the scope for the generation of scientific knowledge. In this way of thinking there was little room for reflexive articulation between science and society. Rather, where such articulation did occur (and some articulation was considered to be inevitable and even legitimate), it was relegated to the outer institutional margins. For example, in Germany the nineteenth century the demand for technically and scientifically trained people was typically relegated to Technical High Schools, which allowed the universities to retain their monopoly on what was termed 'pure' science. The reward structures of discipline-based science in universities seemed for a long time to be designed explicitly to discourage articulation. Even today within the disciplinary hierarchy there is presumed to be a 'hard core' that has resisted the demands of articulation, on the grounds that this would restrict the open-endedness of science, compromise its methods, undermine its creativity and – ultimately – reduce its effectiveness.

But, in practice, scientists do respond to new incentives offered by different 'games' or by changes in the rules of the scientific 'game' – and not simply out of opportunism. Different organizational arrangements may amount to changes in rules because they introduce new players, modify the available and relevant information or shift the rewards associated with certain outcomes. However, the generation of contextualized knowledge does more than reflect changing exogenous conditions; and its production is stimulated not solely by refinements in the organization of research. Contextualized knowledge is also the result of unintentional social experimentation, in which highly industrialized societies have become ever more dependent on the relentless process of innovation. Expectations produced by new socio-economic arrangements, the increasing priority given to industrial research and the growing emphasis on 'wealth creation' gradually seep into the core of scientific activities – even when such activities seem remote from any 'application'. But the imperatives of innovation generally are local, even if the competitive thrust has acquired global dimensions. This is why contextualized knowledge is increasingly produced in, and perhaps primarily relevant to, particular and localized environments and the problems generated within these sites – and why the larger success and deeper implications of contextualized knowledge depend on its spread from one local knowledge-production site to another until the global science system is reached and thoroughly transformed.

Consequently the reflexive articulation of science and society induces a process of continued contextualization of knowledge. The articulation is reflexive because a pervasive discourse is developed which becomes constitutive of the social reality it seeks to portray. Examples can be found in many science-policy documents and official policy statements, in the recommendations of high-level policy committees and, more generally, in the actual transformation of the research and university system which was discussed in the last chapter. University scientists are constantly urged to be more responsive to the needs of industry; academic and industrial scientists are both urged to be more responsive to the needs of users; the market is extolled as a legitimate arena in which scientists have little to fear and much to gain. There is also concern about the need to steer science graduates into professional careers outside the university and research system – for example, into finance or management. But there are also fears of losing the best new university entrants to fields other than science. One strategy is to try to make science more attractive – not only by emphasizing its inherent intellectual sophistication but also by demonstrating its social versatility.

However reluctant researchers may feel initially about being drawn into the wider social arena, which they perceive threatens their dominant influence over the setting of 'scientific' priorities as well as standards of research, there is evidence that the foci of research interests are shifting – what scientists do and how they do it are changing. The transgression of institutional boundaries has become pervasive. The rise of individualism in science – whether in the guise of a new kind of individualistic entrepreneurship or the emergence of new forms of collaboration across collectivist group boundaries – is not only recognized but encouraged. The belief system of scientists is changing as they move from 'segregated' communities into more 'integrated' arenas. Of course, shifts in funding patterns have played an important part in bringing about these changes. Because it is no longer a reasonable expectation to rely solely (or even mainly) on the state to fund research, alternative sources of funding have become more important. They bring with them alternative research priorities, different bureaucratic and policy constraints and novel mentalities and belief systems.

As a result, 'people' – individuals, groups and organizations, whether as consumers or users, as citizens or as addressees of various kinds of policies – increasingly enter the scene of knowledge production. They acquire, as we will show in greater detail in a later section, a place in our knowledge. They demand not only knowledge but also knowledge tailored to their specific requirements and needs. Some are also more adept at mobilizing the expert knowledge they seek and the resources they need. But it would be misleading to conclude that these changes in the scientific culture and in the cultural beliefs of scientists are merely the result of science and technology policy; nor are they produced simply by the gradual, but inexorable, subordination of science to economic imperatives. Both influences, of course, are important. But a more fundamental shift is underway. This larger transformation has been enabled by the general shift from collectivist beliefs inherent in the 'segregation' model towards the more individualistic beliefs and altered patterns of social co-operation and competition characteristic of an 'integration' model. In society at large individualism has been in the ascendant for a long time; in science only recently (and in co-evolution with Mode-2 society) has 'integration', and new forms of co-operation across the science–society divide, become not only possible but desirable.

An Evolutionary Perspective

The rules of engagement are changing because the context is becoming a much more integral part of research – but what actually is the 'context'? So far contextualization has been regarded as a continuous process in which novel forms of interaction and communication take place between scientists on the one hand and on the other those who speak to science. A shift from a 'segregation' to an 'integration' model has led to the opening up of formerly tightly-knit scientific communities, held together with strong collectivist beliefs, practices and social organization. As a result 'users' enter the picture; potential markets matter and funding sources become crucial. Also, public attitudes not only to research products but also to research methods (for example, animal experiments) have become much more influential and demand somehow to be taken into account. New ethical issues and dilemmas arise whenever scientific advance is so rapid that no accepted guidelines are yet in place. The growing power of the mass media continually intervenes, drawing attention to what scientists do and where it may lead. But does all of this really amount to a change that justifies speaking of contextualization of knowledge production? And if so, what is new about it?

Many scientists, of course, may still believe they are engaged in nothing more than a public relations exercise which is inconvenient but necessary. But others have taken full advantage of the opportunities offered by a new entrepreneurial environment. Seen in a longer-term historical perspective the reality-shaping power of science and technology has become a permanent feature of industrially advanced societies, in which material livelihood depends on the ever more effective use of finite, non-renewable natural resources. To be able to advance, or even maintain themselves, these societies must overcome this depletion by achieving an even faster rate of technical and scientific 'progress' – and such progress can only be achieved by the continuous and intensive exploitation of the results of scientific research. This process is accelerating – and being accelerated by – the dynamics of innovation. Once science could claim to have several, perhaps contradictory, functions; today it has become monofunctional (Markus 1987). Its overriding function is to initiate, sustain and be the main driving force behind innovation. Consequently it increasingly forms part of the social reality it shapes. The process of contextualization follows the contexts which multiply.

The diversity of the sciences – some prefer to speak about the 'disunity' of science – has become widely recognized. It has replaced the triple myth of One Science, One Scientist and One Nature. Partly, this has been the result of historical and contemporary social microstudies of scientific practices, of experimental systems, epistemic objects and the like. However, the move towards greater integration, and contextualization, also produces greater diversity within science and between scientific disciplines. Contextualization, by definition, is differentiated and so differentiating. By following the multitude of contexts, it takes on different forms and different languages, with different interlocutors and different outcomes. If context matters overall, so must the multitude of local contexts. But the 'unity' of the sciences is still assured to some extent by existing interconnections between disciplines and specialisms, by the partial utilization of methods and spread of instrumentation and simulation techniques across disciplines and sub-fields and through access to results and methods developed in other branches of knowledge. This 'unity' has to be constantly renewed and reworked. Paradoxically, it is held together by the growing diversity of knowledge produced in different contexts of application which create new spaces of knowledge – dynamic networks of mobile constituents.

Yet, there is a new element which emerges whenever uniformity or homogeneity explodes into diversity and heterogeneity – uncertainty. In chapter 3 its importance in the co-evolution of Mode-2 society and science was emphasized. Following Keynes, uncertainty – as opposed to risk – can be defined as a state in which individual actors find it impossible to attribute a reasonably definite probability to the expected outcome of their choice, as we discussed in chapter 3 (Keynes 1937). In a similar way innovation can be regarded as an experimental process for reducing uncertainty to risk (Schon 1967). In Mode-2 society uncertainty is inherent and a constitutive feature. As a result there is greater (blind?) variation and increased competition; more risks are taken; and many more actors become involved – all of whom seek to attribute a reasonably definite probability to the expected outcome of their choices. This greater uncertainty has positive as well as negative aspects. On the negative side it generates anxieties, promotes retrenchment and paralyses action. On the positive side it encourages greater risk-taking and experimentation. In a situation characterized by uncertainty it would be reasonable to predict the generation of more possibilities and, therefore, of more collaboration and competition. This is what appears to be happening in knowledge production. If it is not clear which direction research will take, what resources will be available and what (and where) leadership will be available, a sensible course of action is to hedge one's bets by keeping options open and participating in several different ventures. In this way a Mode-2 society generates a knowledge environment in which many actors with different skills, perspectives and interests are available for a variety of experimental opportunities – sometimes approaching the limits of stability.

This is very different from the situation which prevailed under the old modernist regime in which Mode-1 science was born and flourished. In a greatly simplified way, we can say that there were firmer institutional boundaries and clearer identities. The role of researchers in knowledge production was shaped by disciplinary norms and professional values. Collectivist beliefs were dominant. Expectations about legitimate interactions and suitable partners were conditioned by membership of particular knowledge communities. Rewards and sanctions were determined by the same communities, which were organized and controlled by meritocratic scientific elites. These elites were responsible for their communities' 'external relations' – negotiations with society and, in particular, with its official twin representatives, State and Industry. The forms in which 'the context was speaking' were often characterized by bureaucratic planning – a top-down approach with articulated aims and objectives, clear lines of responsibility and, for scientists, public funding (with, initially at least, few conditions). Although funding crises did occur, scientists worked within established routines. They did not need to hedge their bets or spread their risks. There were few incentives to enter problem-solving configurations other than those officially designed as either 'pure' or 'applied' research. Because the knowledge production system was more segregated there was less variation among scientific disciplines, less experimentation with novel forms of problem-solving, less interest in collaboration outside fixed scientific communities, and little effort was made to attract funding for research from non-traditional sources.

In a Mode-2 society there has been a shift in the mode of knowledge production from planning to a more open evolutionary pattern in which there are many actors, and resources are no longer fixed and predictable. The management of research has become correspondingly more difficult. Research now has to address a variety of socio-economic objectives and, as a result, research priorities are constantly shifting. It was much easier in the past to implement predetermined goals generated by the triadic regime of Science–State–Industry (with elite scientists influential in all arenas). In response to this greatly increased complexity Mode-2 research has to be more bottom-up than top-down. Elites have also multiplied and diversified. Control is now exercised through a plurality of disciplines and institutions. New research activities arise in cognitive territory populated by many different kinds of experts who possess different kinds of skills and experience. These new activities are also generated in a social context where there is a strong belief that research is indispensable to solving all kinds of problems which extend far beyond those considered to be 'scientifically' interesting. How public- and private-sector bureaucracies used to work and, in particular, how they generated research programmes was well understood. In contrast the emergence and organization of Mode-2 research are poorly understood, let alone efficiently managed, phenomena. But the general increase in uncertainty can be said to be one of the driving forces behind contextualization.

So how does Mode-2 research produce contextualized knowledge? How do novel research configurations emerge from the vast range of possibilities that Mode-2 society presents? The first answer we gave was in terms of a shift from a 'segregation' to an 'integration' model. A second answer emerged through the increasing prevalence and importance of uncertainty. The 'how' of contextualization can also take place through the generation of unprecedented variation and the operation of various selection and retention filters. In the molecular biology laboratories observed by Knorr Cetina, blind variation and natural selection were strategies adopted when confronted with open problems. Molecular biologists vary the procedure that produced the problem, and then let something like its fitness, its success in yielding effective results, decide the fate of the experimental reaction. In this case, variation is 'blind' in a precise sense: it is not based on the kind of scientific understanding and investigation of a problem that characterized, for instance, how high energy physicists work. Molecular biologists do not embark on an investigative journey whose sole purpose it is to understand the problem. Instead, they will try several variations, in the belief that these will result in workable evidence (Knorr Cetina 1999: 91).

In an analogous way, the process of contextualization of knowledge can be said to rely heavily on 'blind' variation – in order to find out what 'works'. This is not to say that the strategy of variation is 'blind' in the undirected and undiscerning sense in which, for instance, genetic mutations occur (from which the term is borrowed). Strategies are never arbitrary, nor are they based on truly blind trials and errors. They are never just odd random alterations either. Experience counts – as does a sense for what could be a 'promising' strategy, leading to success. Informed guesses as to what may have happened or what may work out are just as important – in the world of knowledge production just as in playing the financial markets. But there is not the same urgency of seeking to clarify what the problem really is. Often there are time constraints as well, reinforcing the pragmatic attitude and reliance on 'approximation' which is adopted. But ultimately, the mechanism for obtaining results is strategic (not blind, but perhaps half-blind?) variation and selective retention that comes through success.

The contrast with the planning mode typical of the engagement between Mode-1 science and society when not only goals, but routes and even methods are laid down in advance, is striking (as is the similarity to the meticulous preparation of high energy physics experiments years in advance). Of course, Mode-2 knowledge production is not free from all planning foresight, nor from bureaucratic controls. But an important shift in the regime of control has taken place. Instead of being exercised directly but from 'outside', control is now exercised indirectly and from 'inside'. It is becoming internalized through mechanisms that also characterize the so-called Audit Society. These mechanisms include ever more elaborate systems of peer review, more formal quality control systems and other forms of audit, assessment and evaluation designed to police the consequences of the greater variation of research potential and practice which has been generated by increasing competition. But these practices also create spaces which leave much room for different kinds of (half-blind) strategies of variation to be pursued and many more possibilities in which selection through what counts as 'success' – defined by different kinds of actors and under varying circumstances – can be achieved. These are the many tiny, and not so tiny, cracks in the fabric of scientific knowledge production through which contextualization enters. Once it has entered, there are many more actors now involved, all of whom feel able to judge the acuity of problem-selection and the subtlety of possible problem-solutions. They all feel entitled to assess the 'quality' of research practice, raising thereby a real danger that the new control regimes of the Audit Society, although more diverse, may spawn what Marilyn Strathern has called a new 'tyranny of transparency' which is just as repressive to true scientific creativity and innovation as the old centralized planning bureaucracy (Strathern 2000). If too much emphasis is placed on problem-solutions, the result can be routine research, especially if problems are narrowly predefined, whether by vested interests in industry or by broad alliances of interest and pressure groups. The traditional argument that scientists need to be free to pursue their own research interest – or, if not free, have a dominant voice in the selection of research topics – cannot be lightly dismissed. Autonomy and independence have to be recognized as vital preconditions for scientific creativity – as, indeed, they are for all forms of human creativity.

The key question is how high-quality research can be pursued in the complex situations created by a Mode-2 society. There is a high degree of uncertainty; there is no clear-cut direction but many competing ideas, theories and methods; and no one is in overall charge. The fundamental shift of perspective from advanced planning to evolutionary increase of (half-blind) variation and selective retention implies that the direction taken by research cannot be determined in advance but only *en route*. Everybody, 'producers' and 'users' of knowledge alike, must contend with much higher levels of uncertainty. There is no longer only one scientifically 'correct' way, if there ever was only one, especially when – as is the case, for instance, with mapping the human genome – constraints of cost-efficiency and of time limits must be taken into account. There certainly is not only one scientifically 'correct' way to discover an effective vaccine against AIDS or only one 'correct' design configuration to solve problems in a particular industry. Instead choices emerge in the course of a project because of many different factors, scientific, economic, political and even cultural. These choices then suggest further choices in a dynamic and interactive process, opening the way for strategies of variation upon whose further development ultimately the selection through success will decide.

Uncertainty, so characteristic of a Mode-2 society, promotes competition. Consequent experimentation generates variation. A great deal of activity is the result, even if the formal organizational patterns may not be so apparent. However loosely and informally, this activity is structured by beliefs about the *potential* of particular kinds of science and about its likely implications. Scientists have always had a general sense of when problems are timely and when research fields are ripe for development, even if they have not agreed about the precise details. And they often felt justified in the past in disregarding the consequences of their activity, the use or misuse to which scientific and technological knowledge was put. But in a Mode-2 society others, too, acquire this sense of potential, shortening drastically the link between knowledge and action, between expert advice and putting policies into practice. This is the place for people to enter into our knowledge. As a result, pervasive beliefs, or visions, of that potential push in particular directions where, it is hoped, a more liveable society and a sustainable environment will be created; success is in waiting, and solutions to urgent problems can be found.

The Place of People in our Knowledge

Of course, people are always present in our knowledge, be it as the objects of study and research or as those who are active in doing the research. But it was largely due to the insistence of the feminist movement that this distinction gained in relevance. It is now much more widely accepted that the human objects of research – be they gendered or in other ways socially categorized – have to be carefully attended to, if research results are not to be unnecessarily restricted in their validity or wilfully distorted in their applicability. In clinical research especially, topics like 'gender differences in metabolism' (or in other biological functions) have opened up new and promising avenues for research. But the argument about making space explicitly for people in our knowledge, although it started as a normative issue, can be put on much more general ground.

Conventional arguments about a context-free and universal science and its objectivity are unlikely to have much appeal in the future. As we have argued in the previous chapter, the basis on which the authority of science rests becomes much more closely tied to concrete practices, their results and impacts. Reliable knowledge, although it will remain a solid and indispensable criterion to strive for, will be tested not in the abstract, but in very concrete and local circumstances. If science is to avoid becoming stuck in the objectivity trap, it has to develop greater context-sensitivity. The reliability of scientific knowledge needs to be complemented and strengthened by becoming also socially robust. Hence, context-sensitivity must be heightened and its awareness must be spread. The necessary changes pertain to the ways in which problems are perceived, defined and prioritized, which has implications for the ways in which scientific activities are organized.

One way to make science more context-sensitive is to bring in people. What is the place of people in scientific knowledge? Such a question might first seem to address itself to the social sciences and the humanities, or, already extending slightly further, to 'the social sciences bridge' which is intended to lead to greater interdisciplinary efforts, especially in view of enlightened policy-making (Gago 1998). The social sciences became institutionalized at the turn of the twentieth century, but an institutionalized dialogue between the natural sciences and the social sciences has become possible only in recent years. The social sciences initially were confronted with the problem of how to understand and help maintain a viable social order that would allow the integration of 'the masses' in the process of industrialization, modernization and democratization. Despite enormous conflicts and disruptions, these integration processes proceeded on the political level where the concept of citizenship gradually and, against much resistance, gained acceptance in its inclusivity, both in terms of the social groups who were admitted and in terms of the rights and obligations the concept of citizenship came to embrace. Economically, industrial workers – in themselves a rather recent social category – became integrated in the system of mass production and later in the system of mass consumption. Cultural integration proceeded mainly through the expansion of an educational system which also served to homogenize the national identities of those living within the territorial boundaries of the state. The rising welfare state had to provide mechanisms for warding off the uncertainties and vulnerabilities that came with the disruptive effects of the industrialization (and rapid urbanization) process. At the turn of the twenty-first century, even in the highly industrialized countries, some of these problems are still with us, but their nature and scale has been utterly transformed. The pressing problems of today are seen as intimately related to the emerging global (dis)order, migration and how to accommodate multi-ethnic and multicultural groups in what no longer resembles the idealized image of a homogeneous nation-state and its citizens. The issues of work and employment have taken on central political importance as well as affecting the lives of individual citizens. They are seen as linked to technological change and new demands for skills and qualifications. Questions of life-style and -choices, the future of learning and the information society, as well as the shift from state to market all seem to challenge the received wisdom of how social order is generated and maintained. The concept of self-organization, so fertile and precise in the sciences, appears also at work in wide areas of a Mode-2 society. What first appears as disorder, fragmentation and disunity may well, at a deeper level, reveal new patterns of order, while raising the question of how to manage it, if it can no longer be controlled.

What then is the place of people in the process of knowledge production? Do people appear as passive subjects or as active agents? Are they represented at the micro-level or at the macro-level? The following matrix (figure 7.1) was originally developed to show how and where 'people' were conceptualized in the context of problems of urbanization and sustainability (Nowotny 1999: 257). It can easily be extended to other fields of enquiry. It is important to recognize that all questions and hypotheses represented in the knowledge matrix are legitimate research questions. But what the knowledge matrix also intends to show is that the implications and policy recommendations vary not only with the questions asked (and the motives behind any such research agenda), but with the place assigned to human agents.

But not all fields of knowledge readily lend themselves to include people, although all knowledge will in the end be somehow related to people. In order to strengthen the context-sensitivity of scientific knowledge also in those fields, it is important to recognize how much context-sensitivity already exists. Today, there is a genuine urge to participate in a democratically responsive manner in scientific

Figure 7.1 The knowledge matrix

	Commitment of participants		
Issues emerging at the		Active	Passive
	Micro-level		
	Macro-level		

and technological developments that carry far-reaching consequences for individuals and society alike. There is a shift from a culture of scientific autonomy to a culture of accountability, although the latter is still too reactive and in danger of being interpreted in a formalistic and bureaucratized way. But just as the universality of the nation-state turned out to be an illusion – despite its spread throughout the world – so has it been impossible for science and technology to maintain the social function of providing consensus within liberal democracies. For better or worse, in the eyes of many citizens science and technology are equated with the products they deliver and the results that are attributed to them. They have become equated to these goods, access to which should become democratically regulated and allotment of which should also be fairly distributed. What should be produced, and how, especially in view of potential risks, is therefore also seen to underlie the rules of democratic decision-making. It is difficult to see how, in the face of this widespread utilitarian-instrumental relationship to science and technology, reinforced on the part of political leaders who value science for its value-adding capacity only, the contextualization of knowledge which operates through the vociferous presence of 'people' will diminish.

Conclusion

In the following chapters different forms of contextualization will be discussed. Context influences theories and practices of science in many different ways – which is hardly surprising in the light of the great variety of the research system. There are many constraints on the research agenda – theoretical and epistemological constraints flowing from the definition of problems and choice of research methods, and also practical constraints imposed by collaborative frameworks or access to instrumentation. By analysing these constraints it is possible to get a better sense of the interaction between constraints and context. To the extent that context has been taken seriously, it has been regarded as something 'external' that influences and, therefore, evokes a 'response' from science. Today the argument has to be turned on its head. It is necessary to identify the constraints – or resistances – that shape the process of permanent interaction between context and science in the conduct of research. Constraints are not the same thing as context. Potential is limited by both social and cognitive factors some of which may only emerge in the course of interaction. Contextualization does not imply a dichotomy between science and society, inside and outside. Rather it suggests a spectrum of complex interactions between potential and use, constraints and stimulants. Consequently, different forms of contextualization can be observed: both weakly and strongly contextualized knowledge and also examples of middle-ground contextualization. These examples will be discussed in the next three chapters.

8

Weakly Contextualized Knowledge

Based on the three distinctions made in the previous chapter as to how contextualization happens – through the shift from a 'segregation' to an 'integration' model, through the increase in uncertainty and more variation and selective retention that accompanies it, and through greater awareness of where is the place of 'people' in our knowledge – we are now ready to consider various forms, or degrees, of contextualization. In this chapter and the next a distinction is drawn between weakly and strongly contextualized knowledge production. Two examples of each are discussed. In the case of weak contextualization one example is specific – particle physics – while the other considers national R&D programmes in general (which may come as a surprise to readers because such programmes are often regarded as attempts to shape scientific priorities to fit political, social and economic agendas). In the case of strong contextualization both examples are specific – the Central Artery/Tunnel Project in Boston and co-operation between medical research and patients' self-help groups.

All knowledge, scientific or otherwise, is produced within a particular culture and set of social arrangements. It is not produced at some remote ideal site and then transferred to 'society' to be adapted or shaped to some practical purpose. In that sense all knowledge is contextualized. It is produced by groups of individuals who, however independent their ideas, form a loose intellectual collective which operates in a specific historical context. By emphasizing the non-particularistic character of valid and reliable scientific knowledge the collective enables a relatively stable consensus to be formed. Its members are bound together by shared intellectual orientations, values and perceptions; in most cases they also have common material interests. Nevertheless, for nearly a century science in advanced industrial countries has cultivated an image, and developed an ethos, of separation from its surrounding society. Its fear was 'contamination' by special interests and ideologies considered to be alien to the pursuit of 'pure' science. Thus, the divide separating Nature from Society had itself become an ideological component.

At the start of modern science, social distance was used to create a social space of relative autonomy from traditional value systems and practices, notably received religion, and from powerful links to established authorities. Without such a distinct and protected space, modern science would hardly have taken the institutional routes that enabled it to flourish, as historical and comparative research shows clearly. Later, as science became firmly institutionalized inside universities in the nineteenth century, it aligned itself with the new and self-confident university ethos and placed a higher value on knowledge-for-its-own-sake, relegating 'applied science' to industrial laboratories. University-based scientists treasured the freedom to pursue research, wherever it might lead and (almost) regardless of consequences. This orientation towards curiosity-driven science has often been interpreted as indifference to a society's urgent need to find answers to pressing problems. Of course, since the time of Bacon and Galileo science has always aimed to be useful. But this utility, it was argued, could only be realized if scientists were free to set their own agendas unhindered by political and/or market considerations. This formulation of scientific freedom supported the idea that scientists themselves were the best judges of the problems that had to be addressed and was the guiding factor in the development of research agendas until the middle of the twentieth century.

In the immediate period after the Second World War, in a very early phase of science policy, a crude balance was struck. On one hand was the belief that problem-choice was indeed the prerogative of scientists; on the other was the recognition that problem-choice needed to accommodate social demands for utility. Alvin Weinberg, in an influential paper, distinguished between 'internal' and 'external' selection criteria in choosing between different scientific projects (Weinberg 1963: 159–71). 'Internal' criteria reflected the structure of a particular field, the quality of available manpower, the extent to which a scientific field was ripe for further development and its relationships with developments in other fields. These criteria were matters to be judged solely by scientists. In contrast, 'external' criteria – the social, economic, political and cultural benefits of pursuing a particular line of research – were left to the judgement and decision-making of 'external' experts (whether from industry, government or society at large).

According to Weinberg, a strict sequence of decision-making had to be observed. First, decisions about funding research projects were taken by scientists solely on the basis of intrinsic scientific merit; then the social and economic merits of particular projects were determined by a wider community of interests on the basis of a diversity of social, economic and political considerations. For Weinberg (and for the generation of scientists and policy-makers influenced by his ideas), science policy should attempt to insulate assessments of scientific quality from, if not its potential use, its potential users. Weinberg did not argue that only truly excellent research should be supported, but that consideration of the potential social benefits of research should supplement, not supersede, the already established case for supporting research on its intrinsic scientific merits. The choice criteria preserved the paramount role of scientists in making scientific judgements about worthwhile research. Thus, decisions were to be taken in a clear sequence taking into account two different considerations – first, the intrinsic merits of scientific projects; and, second, external criteria which typically related to socio-economic utility. It was also clear that these considerations had to be evaluated by different groups.

Weinberg's criteria were originally used to help the state, and its agencies, choose between large, and inevitably expensive, scientific projects – for example, a particle accelerator or a telescope. But the application of 'internal' and 'external' criteria, even to fundamental science, explicitly acknowledged the existence of a broader social context. Because scientists now had to compete for funding, constraints appeared, not only dictated by nature (which was easy to accept) but also imposed by society (which was less congenial). For the first time scientists, even those working at the cutting edge of fundamental research, had to explain to others, including other scientists, the potential contribution of their field both to the development of science and to broader social objectives. How far this early formulation of limited contextualization influenced any decision-making processes (beyond the rhetorical level) is unclear. But common sense suggests that, because of increasing resource constraints, many debates were conducted along lines similar to those which the Weinberg criteria were trying to make explicit.

Particle Physics

Is it fair, however, to regard the science generated in, for example, the CERN, Brookhaven, and Hamburg laboratories as contextualized knowledge simply because decision-making was guided by something like the Weinberg criteria? In the sense that physicists have gradually had to explain the potential benefits of particle physics in terms that go beyond narrow disciplinary interests, the answer is yes. As socio-economic constraints became more evident, particle physicists began to produce evidence that their research would also provide manpower for industry and strengthen national technological capability through the production of new instrumentation and equipment, and even suggested the possibility that particle physics could provide a basis for nuclear power based upon the energy that bound not the nucleus but the nucleons themselves together.

It is clear from the history of these projects that there was a dialogue between the scientists and the (mainly government) funders, as there was between the machine-builders and industry (Krige and Pestre 1997). Of course, in countries with little in the way of high-technology industries such work could never be funded. But in advanced industrial societies – societies, moreover, which were engaged in a technology-driven Cold War with the Soviet Union – the political context loomed large in the decisions that led to the building of internationally funded, collaboratively run large particle accelerators. So there was certainly a context – but the linkages between the context and scientific content were tenuous, although powerfully mediated through military and political concerns (Krige 1996). There was vigorous and extremely fertile interaction going on within the tightly knit particle physics community, which contrasts strongly with the lack of feedback between 'producers', particle physicists, and 'users' not simply in government and industry but also scientists from other disciplines. In this case the development of the research agenda can be explained primarily in terms of scientific considerations, almost exclusively shaped and run by an exceptionally cohesive community of scientists. They conform, as we have shown in the previous chapter, almost ideally to the 'segregated' model of social organization generalized by Avner Greif. A growing awareness of cost was an important consideration but one which came only as secondary. For example, there were times where concern about escalating costs prompted particle physicists to look for less expensive design configurations. Viewed in this perspective, particle physics could never have been more than an example of weakly contextualized knowledge.

There is a further reason for this. Particle physics is an area where collectivist beliefs and behaviour are still strong in the scientific community. Anthropological studies, such as those conducted by Sharon Traweek, show a tightly knit and inward-looking community with 'tribal' features. Knorr Cetina has described the social structure of the field as a 'super-organism', in which individual members lose their individuality and are merged in the collectivity. Because particle physicists work on experiments with long time-horizons and, therefore, precise planning-horizons, powerful social bonds are forged that find expression, among other things, in the practice of publishing scientific papers with a hundred or more co-authors (Traweek 1988; Knorr Cetina 1999). As a result there is little room in particle physics, in terms of its scientific content and tightly interwoven collaborative practices (inevitably centred round a highly specific site and a big machine), forms of work, epistemic culture and social attitudes, for a loosening-up of social relations associated with a shift towards a model of 'integration', or the adoption of working methods prevalent among the life sciences.

Since particle physics tries to understand the basic components of the universe, there is also no place for 'people' in its knowledge (although people enter, of course, in the 'communitarian' framework as colleagues, students, 'other groups' and what Knorr Cetina calls 'gossip circles'). In this connection it is tempting to speculate about the anthropomorphic features that the detector acquires. In Knorr Cetina's description, detectors take on individualized idiosyncrasies; each being 'different' from another, with physiological characteristics ascribed to them, such as ageing, diseases and life expectancy. Detectors are also treated in a way that makes them analogous to becoming social and moral beings. They co-operate; communicate with each other and with other devices; they may consult and check each other. They behave or misbehave, can be trusted or are to be distrusted, and are held responsible for certain achievements (Knorr Cetina 1999: 119–20). And Carlo Rubbia, one of the great detector designers, is quoted as saying: 'Detectors are really the way to express yourself. To say somehow what you have in your guts. In the case of painters, it is painting. In the case of sculptors, it's sculpture. In the case of experimental physics, it's detectors. The detector is the image of the guy who designed it' (quoted in Galison 1997: xviii).

But however 'people' (or a particular individual, the designer) may enter the field of particle physics which is seemingly so devoid of 'people', this is not the end of the story. However weakly contextualized particle physics was initially, the debate about the consequences – and, inevitably, cost – of increasing technical sophistication that began with Weinberg continued over two decades. By then, the overall societal context had changed in decisive and irreversible ways, last but not least with the end of the Cold War marking the beginning of a new era. Thus, a number of factors came together – the upward cost trend, greater awareness of the public accountability of all scientific projects funded by the state and disagreements with other disciplines about the wider significance of particle physics put the subject under increasing pressure to set research priorities.

Perhaps for the first time in the history of physics, open dissension surfaced among physicists in public when influential solid-state physicists testified in congressional hearings against their colleagues that they had failed to make their case. The result of these and other factors was the unprecedented decision of the United States Congress not to fund the super-conducting super-collider, although substantial funds had already been committed and the mega-project was well on its way. The political view was that 'enough is enough', and particle physicists were urged to explore the possibilities of international collaboration. In this case there was little convincing articulation of scientific problems in relation to the needs of other physics specialisms (notably solid-state physics) and to those of other disciplines (such as biological sciences) or to the broader socio-economic context. This is now changing – but too late. The failure to achieve a broader consensus about the importance of the problems pursued by particle physicists or to identify other uses to which such an accelerator might be put led to an over-concentration on the funding issue. In effect, a scientific mono-culture had been created. Like all monocultures it was vulnerable to changes in 'market' conditions. In a national context the super-conducting super-collider collapsed, although it has re-emerged in the international context, under the auspices of the Megascience Forum convened by the Organization for Economic Co-operation and Development (OECD). To adopt a phrase that will be explained in chapter 10, particle accelerators failed to function as 'Mode-2 objects' for a wider range of scientific and social interests.

The example of particle physics reveals the many layers of context. At the most general level particle physics was associated with the general defence aims of the West, particularly the United States. First, the development of particle physics cannot be separated from the development of nuclear science more generally, which in turn cannot be entirely separated from the technological imperatives of the Cold War. Indeed one reason for the eventual failure of the super-conducting super-collider is that Cold War imperatives were beginning to wane when the project came up for approval. Although it would be wrong to say that the Cold War directly shaped the development of particle physics, equally, without that overriding concern in the West with security, the field would have developed more slowly. Second, economic factors were already beginning to influence the research environment even when Cold War pressures were still strong in the 1960s and 1970s. Cost constraints were forcing engineers to redesign particle accelerators. Inevitably the size and power of available accelerators shaped the kind of physics that could be done. Physicists learned very quickly to adjust their questions to the performance characteristics of the machine that could be afforded. The outputs of particle physics have been conditioned by political and economic factors – but they have also been influenced by the design capabilities of industry (for example, in the technology of large magnets, vacuum systems and large bubble chambers).

There are still more dimensions of contextualization. The physical location of accelerators was determined by geography and politics. As a result project planners were drawn into specific regional economics and local politics. So, even in a field apparently as remote from utility as particle physics, a significant degree of contextualization is evident. Various constraints have affected the actual performance characteristics of accelerators and, therefore, the research agenda which could be pursued. Furthermore, particle physics, despite the powerful scientific and technical constraints that shaped it, also had to take account of constraints emerging from many layers of soft context – global political considerations, national economic strength, industrial capability, and local and regional geography. Particle physics, therefore, did produce contextualized knowledge – but weakly contextualized knowledge in the sense that the social context provides only a weak signal about shaping the research agenda. But the ability of particle physicists to alter the context had also become weak. Once they had lost their powerful status and influence which resulted from their triumphant engagement in the War effort and which had been prolonged by the security concerns of the Cold War era, they apparently had few other resources left that could now usefully be employed in the new game over local politics, national funding patterns and the general demand for greater accountability.

National R&D Programmes

National R&D programmes provide other examples of weak contextualization. Within such national programmes research priorities are identified within a structured framework and financial resources earmarked to support these priorities. Scientists then 'respond' to the research opportunities that these programmes offer. Sometimes they change research direction, but more often they adapt work already in progress to match the aims and objectives of national programmes. Often these programmes involve joint funding between governments, industry and universities. Their (unrealizable?) aim is to ensure that, as far as possible, the benefits of research and innovation are captured within national boundaries. But, within these national contexts, scientists still determine the agenda; typically they play a dominant role in initiating and managing national programmes. The dominant context is the nation-state. It sometimes speaks with one voice – the state; sometimes with two – the state and industry; and occasionally with three – state, industry and the basic research which is performed inside universities. The requirements and constraints imposed by society are predictable, planned and usually co-ordinated and managed in a centralized, often bureaucratic way. As a result, the messages from 'context' to 'science' are usually very general – for example, the exploitation of important new technologies to produce greater industrial productivity or to pursue mission statements about 'key technologies' which are usually identical across the industrialized nations. To a considerable extent these messages can be ignored or re-interpreted by the research communities which contribute to these national programmes (Küppers, Lundreen and Weingart 1978; Weingart 1997). Again, contextualized knowledge is certainly being produced – but only weakly contextualized because research agendas are still relatively open and under the influence of the researchers themselves (although not as open as research funded by research councils or charitable foundations). For the major part, national R&D programmes are not properly reflexive; nor, as yet, do they sufficiently promote further 'integration'.

NASA-type projects go one step further. Research is within a tight hierarchical framework from the start and most of the research is structured according to specific technical performance requirements. Messages from 'outside' the context – for example, cost considerations, the environmental impact of rocket launches or considerations of other, competing research priorities – are either translated into technical tasks, which then may or may not be supported by research components, or are ignored entirely. Scientists in NASA-type organizations are not expected to wander from mission objectives. The research context is rule-dominated, which can lead to disasters such as were experienced in the later Challenger missions (Vaughan 1996). In a word, in NASA-type programmes research is constrained by the organizational environment (and, in that narrow sense, is contextualized). But, because this environment is tightly structured, the context can speak to research only in very limited respects. The knowledge produced in this type of programme remains only weakly contextualized, however powerful the other organizational constraints alongside scientific and technical ones may be. Such constraints, and their tension-ridden interaction, were also at work in the Manhattan Project, with the military intervening repeatedly at decisive points (Kevles 1987).

Weak contextualization also characterizes other government-supported research programmes. This applies both to programmes designed to meet specific socially mandated objectives, such as the spate of former 'wars on cancer', and to those with more general goals – for example, to develop information and communication or nano-technologies upon which future competitiveness is thought to depend. These programmes are critically dependent upon the 'response' of scientists in both universities and other knowledge-producing institutions. So it is often difficult to demonstrate that programmatic research differs significantly from open-agenda research, or whether, in practice, such research is related to programme goals. There is often limited dialogue between those who formulate the programme objectives and those who put in proposals for research projects or grants. Typically programmes are too bureaucratically structured for such a dialogue to take place. It is hardly surprising that scientists play to their research strengths and try to modify their research interests to accommodate the objectives of the new programme. As a result, many government research programmes produce only weakly contextualized knowledge, despite the clear intention of relating research to social and economic objectives. The Fifth Framework Programme contains an explicit attempt to move closer to users and user needs. In its rhetoric and political intention it aims at developing strongly contextualized knowledge, but it remains to be seen whether the heavy-handed bureaucratic structures, both within the European Commission and among member states, will be able to achieve this goal.

Bureaucratic management of research is only one reason for the production of weakly contextualized knowledge. Another is an (understandable) reluctance of researchers to get more deeply involved with social context. There are now clear signs of movement away from bureaucratic research management to a new culture of programmatic research which places greater emphasis on links with industry (although this may simply mean that industry's more short-term priorities, based on market demands, are being adopted). The bureaucratic approach of national programmes worked well when the funding for science came primarily from government and when research funding was legitimated by links to 'national' objectives. Both as a framework in which research priorities could be developed and as a mechanism to fund these priorities, bureaucratic research management was appropriate for science under conditions of what still was largely 'segregation'. The changes that have been under way recently in the move towards more 'integration' are also experiments in coming to terms with the complex cognitive landscape of a Mode-2 society.

In retrospect, it is not difficult to see why the knowledge produced was only weakly contextualized. First, whatever discussions there were among the scientific community before setting up a national programme, the need to move closer to the social context was not a high priority. It was the government's, not the researchers', responsibility to translate social need into research programmes or vice versa. Second, power and influence were concentrated in the higher levels of government, because governments were expected to pay for these programmes. Politicians – and, more often, civil servants – had to re-interpret social needs to fit the parameters of these programmes. Third, structures based on traditional bureaucratic models were established and resources flowed along channels that were difficult to re-route. Fourth, unsurprisingly, scientists found that provided they did 'good science' they could leave concerns about the extra-scientific impact of their work to those higher up the national programme hierarchy.

To summarize, science has always been carried out in a social context. But a regime of 'segregation' established a clear division of labour and responsibilities between scientists, governments and industry. It discouraged idiosyncratic and not centrally controlled moves across these boundaries. So the context was only able to send out indistinct or weak signals. University science, in particular, could protect itself from too close an interaction with society by ensuring that the conduct of research, whether in the choice of topic or of method, was a matter to be settled by scientists themselves. Of course, in maintaining control the guardians of the institutions of science were only partly successful. Science was regarded as a public good and had to rely on the support of public funds. As governments became more aware of the costs of research they sought to direct funding to particular objectives. However, despite the complaints of many university scientists, it is not clear that the growth in the number of government-funded research programmes has really made research more context-sensitive. As a result, much research deliberately directed towards social and economic objectives still produces only weakly contextualized knowledge.

9

Strongly Contextualized Knowledge

In the previous chapter we attempted to outline the characteristics of weakly contextualized knowledge, and offered two case studies. One case study involved a set of bureaucratically managed research programmes. Such programmes are sometimes portrayed as examples of the subordination of science to social goals and, therefore, ought to be regarded as examples of strong contextualization. But the imposition of social imperatives on science is not what is meant by contextualization. By contrast, strong contextualization occurs when researchers have the opportunity, *and are willing*, to respond to signals received from society. It is a dynamic, two-way, process of communication, the very antithesis of a process which seeks to control science through bureaucratic means.

When strongly contextualized knowledge is produced – as is the case when one or all three of the conditions spelled out in the previous chapter are met – either beliefs and their legitimacy are changed in the direction of stronger interaction with 'outsiders'; or uncertainty is increased and so is variation and selective retention through 'success'; or 'people' have entered the research process in manifold guises, be it as consciously reflected 'objects' of research (for example, when clinical trials are extended to deliberately include women) or as actors whose needs, wishes and desires are listened to, and possibly responded to, if not anticipated. It is important to emphasize that strong contextualization not only shapes research agendas and priorities, but also influences research topics and methods. It enters into the process of knowledge production and therefore leaves visible traces in 'the science' itself. It does so by altering the perspectives of the researchers – and so their definitions of problems and of what constitutes valuable research. Strong contextualization does not establish an authoritative set of aims, objectives and projects to be pursued. It does not tell researchers what they must do. Instead, strong contextualization thrives on communication, a good deal of opportunism and opportunities, and continued interaction which, if prolonged sufficiently, may lead to new approaches or to the definition of new problem-areas. It is in the nature of such interactions that the beliefs of policy-makers and the public may be shaped by scientists, but the latter are also not immune to the projections of scientific and technological advance which policy-makers may have. These too can energize and give focus to research that scientists wish to undertake.

In a sense, the environmental sciences which burst upon the research scene in the wake of the influential environmental movements of the 1970s can be said to be an exemplar of strong contextualization. But part of their research history also accords nicely with what political scientists call the 'garbage can model' of policy-making. In this model, science policy and politics are loosely coupled and change is discontinuous. Because the scientific agenda and the political strategies of the scientific elites are not connected, the scientific agenda, rather than growing from scientific findings, may shift or accelerate, often unpredictably, from political developments in previously unrelated policy arenas. Entrepreneurship – or opportunism – by leading members of scientific elites, thus serves as a mechanism by which these developments influence the scientific agenda. Because the research carried out in this way often lacks a clear relationship to the environmental policy agenda, the results are likewise likely to be relegated to the 'garbage can', interpreted in this model as an incoherent assemblage of scientific results (Hart and Victor 1993: 668). In the 'garbage can model' research lacks 'integration' with the policy agenda. For the most part, however, in environmental concerns, 'integration' of research and public debate are strongly present and pervasive, as evidenced by the long series of international conferences concerned with the environment, from Stockholm to Rio, Kyoto and Buenos Aires, where government scientists mingle with other policy-makers, university scientists and representatives from non-governmental organizations (NGOs). Hybrid committees, made up of scientists, policy-makers and representatives from NGOs, set agendas and debate scientific findings in relation to policy measures that need to be adopted. 'Integration' works on many different levels, from the grass-roots measurement and monitoring of environmental degradation which extends from the local to the national and international level, to the inclusion of representatives from the public in novel forms of experimental consensus conferences or of lay people on the boards of environmental research institutions.

It might be argued that in the case of the environmental sciences, 'integration' is not so much a matter of free choice, but one of necessity (and/or scientific opportunism). The complex inter-linkages between the natural environment and human intervention leading either to its degradation or, hopefully, to better prospects of its sustainability in the future, require thoughtful consideration of the links between knowledge and action and of science, politics and policies, and following up the implementation of the latter. The constant necessity to negotiate – again, on different levels, with different kinds of actors, in different kinds of political and economic arenas – has also encouraged a working method which relies on voluntary and involuntary 'experimentation', guided by strategies of (half-blind) variation while anxiously waiting for 'success' to act as a selective filter. This includes the working of large-scale climate modelling, where the jury is still out on the predictive ability of these models. Uncertainty in the environmental sciences is greatly enhanced as a result of the unpredictability of human action as well as of the sheer complexity of interaction between the natural environment and the effects of humans on it. Last, but not least, 'people' are abundant, even overabundant, in more than one sense. Through public awareness they may influence research agendas, but they are also the targets of policy recommendations – often in those parts of the world where there are intractable problems in making their voices heard or even in recognizing what their (environmental) needs are. The grassroots activism of the environmental movement has transformed itself over time into a powerful 'epistemic network' of non-governmental organizations, which now regularly make use of their 'right' to be consulted and to participate in decision-making at international conferences.

Strong contextualization in the case of the environmental sciences produces a very different atmosphere from the one in which scientists pursue their 'private' research agendas under the guise of a national or industrial research programme, or connive in making grant applications for research they have already done in order to use the money to conduct their own chosen research. But this is not a moral issue (and the temptation to continue to pursue one's own research agenda under a publicly approved, and therefore convenient, research label, may still be strong also among environmental scientists). What is at stake is the urgency to take the context seriously, to let it enter into the research that is being undertaken, since otherwise this research would miss its own objectives and goals. The environmental sciences are sometimes put under the rubric of 'policy-driven sciences'. Policies, their implementation and the rationale for undertaking them in the first place may play a greater role than in other fields, but this in no way detracts from the formidable challenges that propel what amounts perhaps to another kind of research agenda: one that seeks to integrate also the natural and the social world and explore how the two might live together – sustainably.

Contextualization, therefore, depends on a permanent dialogue between scientists and diverse 'others' in society. It is multi-layered. Explicit messages are communicated, interpreted and re-interpreted, as well as implicit or yet-to-be articulated preferences, needs and desires. Strong contextualization requires a common understanding about the nature of an issue or problem and of the role of research in dealing with it. In the field of environmental policies, epistemic networks have helped to diffuse national or international environmental objectives and, as a result, have stimulated convergence towards higher standards of performance, although typically through decentralized means and much cross-cultural learning (Yearley 1996). Another feature of strong contextualization is the growing importance attached to the capacity to translate knowledge into action. In this context Jasanoff has observed how freely the processes of scientific fact-making have accommodated themselves to the demands of politics (Jasanoff 1997). If communication, interaction and dialogue are to find concrete expression in forms of research activity, they need a medium or communication forum of some kind which brings potential participants into sufficiently close contact to keep the process moving forward. But 'success' cannot be guaranteed; rather, it will 'select'. There are many reasons why the messages, signals or even pressures from society may not attract the interest of researchers and so fail to be translated into research programmes. All contexts come with constraints. In particular cases these may be strong enough to preclude certain possibilities. In terms of predictability of outcome, Mode-2 research is as uncertain as Mode-1 science.

The Central Artery/Tunnel (CA/T) Project

It is not easy to identify unequivocal examples of strong contextualization, especially when such examples should also demonstrate what a difference strong contextualization makes. We therefore take first one example from technology rather than science. Of course, it is easier to identify the production of strongly contextualized knowledge in a complex technological project. As Thomas Hughes and other historians of technology have shown, the historical unfolding of large technical systems, like the project of electrification in Europe or the US, has often proceeded in stages, in which various kinds of constraints and obstacles – technical, but also economic, political, financial and even legal – emerged and were eventually resolved, only to encounter new kinds of obstacles and constraints. Hughes and his followers have insisted that it does not make much sense to distinguish 'external' constraints from 'internal', or technical, ones. Rather, they should be considered in their entirety as the 'seamless web of society and technology' (Hughes 1983 and 1987).

But the 'momentum' which large technical systems eventually came to acquire in their maturity and their proneness to exact the technical over the social owe much to the age of modernity. They can be interpreted as the very embodiment – even celebration – of the primacy accorded to order and control. It is therefore not surprising that more recent technological systems like 'management systems', and the technically hybrid telematic systems of the information technology age, have come to espouse, and express, a different spirit altogether. Accordingly, Hughes calls these non-grid based, hybrid projects of the 1970s and 1980s 'post-modern' technologies. As a rule, they involve not only technical, or engineering problems, but substantial political, social and environmental issues. But what makes them different from their pre-modern or modern predecessors is how professional engineers approach the problems they encounter. They cope differently with complexity. Compared to the relatively clear national defence goals that drove American military projects like the SAGE air defence system and justified the Advanced Research Projects Agency Network (ARPANET) expenditures, in which a single agency was responsible, the more recent projects display all the features of messy, complex systems. Not only are many agencies, federal, state and local governments involved in these civil projects, so are numerous local interest groups, including those whose intent may be to protect ethnic neighbourhoods and the environment. The physical and engineering design of these projects invariably reflects these interests on various levels. As the example provided by the Central Artery/Tunnel (CA/T) Project in Boston will show, the move towards greater context-sensitivity and the closer involvement of potential users or presumed beneficiaries is a relatively new phenomenon in the design and implementation of large-scale technological projects. It is not, as yet, a widely accepted practice, which makes the lessons provided by it all the more interesting.

The technological transformation that occurred in the period preceding and following the Second World War can itself be read as a shift towards strong contextualization. According to Hughes, the maintenance of a system for mass-producing standardized products was paramount in the work of pre-war modern engineers and managers. On the other hand, the post-war engineers and managers were no longer rigidly committed to standardization. Instead, they tolerated, even embraced, heterogeneity. Managers and engineers in the modern period expected that the problem-solving techniques they had mastered as young professionals would change only marginally. In contrast, post-Second World War project engineers and managers needed to keep continuously abreast of rapid changes in these techniques. While linear growth had been the hallmark of modern management and engineering, discontinuous change is the expectation of project professionals today. 'Modern' professionals assumed research and development to be a prelude to an enduring innovation. In the post-modern era, project engineers and managers realize that managing innovation is the norm, and anticipate a sequence of R&D projects.

A far-reaching result is that standard hierarchies now have to be modified to allow for local initiative. The resulting compromise, called 'black-boxing', allows local R&D teams to choose the technology that will fulfil system specifications. When, for example, systems engineers specify the thrust required from a missile propulsion engine, the propulsion design team has leeway in choosing the mechanical, electrical and chemical means for achieving such specifications. The systems engineers need not look into the 'black-box' within which the team is working, and micro-manage the means used to achieve the specified ends. Pre-Second World War manufacturing firms valued the highly trained engineering specialists who concentrated on mechanical, electrical or chemical problems. By contrast, post-war projects foster an interdisciplinary approach that cultivates the generalist. 'Modern' experts long familiar with a firm's product line expected deference from young engineers and managers. Today, experience counts for less on projects with a large R&D component because the problems encountered are often unprecedented (Hughes 1998).

The Central Artery/Tunnel (CA/T) Project in Boston Harbor has a long and tumultuous history. Many of the leading figures in this story were socialized in the years of the counter-culture movement of the 1960s with its distrust not only of military technology but of all large-scale technology. The CA/T is a huge project, involving 3.7 miles of tunnels, 2.3 miles of bridges (one across the Charles River) and 1.5

miles of surface streets. It was undertaken in the middle of a city where, two decades earlier, opposition to highway building had been exceptionally strong. The plan, then, was to build more highways in Boston, an Inner Belt, but it provoked an outcry of opposition and led to the formation of a political alliance that came to be known as the Anti-Highway Coalition.

The CA/T style of management and engineering was a response to the noisy and effective demonstrations in the 1960s and 1970s against urban highway constructions. In the new style, engineers had to negotiate with hundreds of neighbourhood, business and environmental groups as well as developers and individuals in response to their putative concerns about anticipated negative environmental impacts. Affluent organizations even hired their own engineers to detail alternative designs for highway alignment, ramps and locations of ventilation buildings. Less well-endowed groups mustered political influence in a manner reminiscent of the successful efforts of neighbourhoods almost two decades earlier to block the inroads of highways and to foster rapid transit.

Through public hearings, the project fostered participatory designs. Experts started to pay attention to local citizens and realized that the originally foreseen forced expropriation of residences and businesses could be avoided. (In the 1950s, about 1,000 residential and commercial structures had been expropriated, evoking an outcry among the local citizens affected, many of whom were Italian by ethnic origin. The building of an Inner Belt would have required even more expropriations and the demolition of almost 4,000 houses in a predominantly black neighbourhood.) Eventually the project's management structure was changed. The 'functional organizational structure' which subdivided management into functional departments such as engineering (design), procurement, project services and construction, was replaced by 'an area-responsibility organization'. Overall responsibility was placed on about half a dozen geographical area managers, rather than functional managers, with much improved results in containing costs and meeting time schedules.

The participatory shift in strategy by the CA/T managers and engineers can, of course, be interpreted as an attempt to mitigate the disruption created by all major public works in a project of this size. But this would be utterly wrong. Instead, many see in it a sea-change in the philosophy of mega-projects: a shift from the imposition of such projects on an unwilling community to the inclusion, by means of thorough consultation, of that community in the decision-making process. No longer can technical experts or political leaders presume that they have the right to make such decisions; rather, the community at large must participate in them. This transition is accompanied by the belief that engineering designs are actually improved by public participation. As Kenneth Kruckemeyer put it in a public lecture at MIT in 1995 '...even a thoughtful engineer will not come up with a good design unless he works under constraints, of which public participation is one...It's the learning process back-and-forth that allows a project to become better...the only federal highway projects which won awards for good design are those which have been sued in ways forcing changes in design' (Kruckemeyer 1998: 118).

The extent of this transformation – from imposition to inclusion of the community – can be appreciated by contrasting it not only to attitudes that prevailed among engineers earlier in the century (and which were encoded in Taylor's doctrine of 'scientific management'), but also to practices that still exist elsewhere today. There is a scene in George Eliot's *Middlemarch* where railway surveyors invade the countryside of nineteenth-century Warwickshire that would have been familiar to Shakespeare three centuries earlier – and are chased from the scene. Loren Graham, an expert on the former Soviet Union's science and technology system, reminds us that in Russia and China today, attitudes and practices still persist that are not dissimilar to those recently abandoned in the West, making the contrast all the more stark. In the Boston CA/T Project, in public hearings, community participation and environmental impact statements became an integral part of project design. For example, a tunnel was shifted so that it emerged at Logan Airport instead of in a residential district, and in building it not a single private residence needed to be demolished. In contrast, another great mega-project – the Three Gorges Project in China – is being constructed according to the old model. The management of the project reflects the values of Soviet-style central planning. When asked if the project might not have harmful effects on the environment and on society, its chief engineer replied: 'Everything in the world has both positive and negative factors or influences, not to mention the Three Gorges Project, which is such a big project. Our policy decision is made on a scientific basis, which reflects the fundamental interest of the people' (Graham 1998: 120).

Of course, some care must be exercised in suggesting that the CA/T project is a paradigmatic example of strong contextualization. First, it can be argued that the project is exceptional, because it is rooted in unusual political circumstances which reflected the counter-culture of the 1960s and the local traces it had left in some of the major figures of the case study. But that is the point of contextualization. Its effect is localized and reversible, not generalized and immutable. Second, of course it is possible to characterize the CA/T project as a cynical exercise in 'participatory planning', in which a new style of engineer skilled at manipulating the public with public relations skills has superseded the old-style engineer who impressed the public by claiming superior technical knowledge (Graham 1998: 122). But even such an interpretation does not detract from the CA/T project's value as an example for how mega-projects which reshape the lives of entire communities can also be conducted. In this sense, strong contextualization makes all the difference: 'CA/T is not an elegantly reductionist endeavour; it is a messily complex embracing of contradictions.... CA/T has been socially constructed, not technologically and economically determined' (Hughes 1998: 304).

On our list of conditions favouring strong contextualization, the CA/T project stands out because of its 'messy complexity' and the strategies that have successfully been adopted to cope with it. The project managers showed a new understanding of how technological and social systems work. But the successful strategies also underline that such an understanding does not rest in a void. Rather, it is embodied in a new breed of professional engineers who feel comfortable with engineers and with representatives of other disciplines and professions, both in academia and industry. Experienced in working within large organizations, the project managers know how to use political power and at the same time how to negotiate imaginatively – with environmentalists, downtown business people, professional engineers and architects, residents of ethnic neighbourhoods and the poor. Strong contextualization has also turned out to be an important learning process.

Medical Research and Patients' Movements

Strong social involvement of scientists in social movements has traditionally been another route towards increased contextualization, although it has usually been hampered by two limiting conditions: the oscillating rise and decline of social movements has inevitably led also to ups and downs in such a *rapprochement*, and the impact of such interaction often was rather restricted from its beginning, due to the relatively small numbers of scientists involved. Nevertheless, there have been repeated instances where such resourceful relationships have been established, with obvious benefits for both sides (Blume 1987). Speculating about the notion of 'legitimacy' that scientists accord to such interests, Blume argues that what is essentially involved may be a sense of shared social purpose, a kind of common social project which is envisaged by all parties. Such a commonality might reside in shared values (commitment to health, to regional development, to the emancipation of women), or shared interests (in the sense of mutual benefits) or a shared sense of social structure (Blume 1987: 34).

But while some forms of co-operation have acquired a highly respectable status and are considered desirable as well as feasible by all – as is the case for university–industry co-operation – other 'common social projects' continue their precarious existence on the margins (like the 'science shops' initiatives which sprang up in the late 1970s in several European countries) or have again slipped into oblivion together with the social and political movement that spawned them (for example, the radical science movement). One of the most successful cases has been what originally was termed 'women's studies' – a mixture of self-help groups, feminist movement strategies and research into the past and present conditions of women with the explicit objective to end discrimination and to achieve the betterment of overall conditions for women. Another field which might be said to be in the ascendancy is also heir to a social movement – partly the women's health movement and partly the emancipatory claims of patients for their rights and support of themselves and their families. It is the growing co-operation between medical research and patients' self-help groups.

In the case of research into muscular dystrophy in France the boundary between science and society almost vanished, so intense was the communication between the context and science as reported by Latour (Latour 1997). Although his account of the development of research into muscular dystrophy has been criticized, it highlights some characteristics which can be identified with the initiation of a 'common social project'. Two perspectives developed initially. First, patients suffering from muscular dystrophy were able to identify their common interest and come together for mutual support. Some were convinced that scientific research would enable them to obtain new knowledge which would alleviate their condition. They believed that research had potential to help them. As in other cases aiming at co-operative links, a process was initiated to find the relevant expertise. Second, some researchers in the bio-sciences also concluded that certain lines of scientific inquiry in areas such as biochemistry or genetics could potentially help to develop more effective treatments of certain diseases, including muscular dystrophy.

As a result an effort directed at translating a problem which was acutely felt by the patients suffering the disease and their families into one that was considered 'researchable' by scientists emerged. Through rudimentary communications between patients and researchers, in which they articulated their individual perceptions, their voices at least were heard. In this interaction the shared belief in the potential of research began to be clarified. Various lines of inquiry emerged; some were discarded and others developed. In time, researchers, patients and administrators generated a common vision, and developed a shared mission. Resources were needed. Towards this end a national telethon was organized. Funds were raised, buildings constructed and administrative procedures developed. Finally an environment was created in which patients and researchers were engaged in a collaborative effort to tackle muscular dystrophy. The research agenda evolved with the help of the patients who provided essential data for further research. The role of individuals was crucial. The patients did not wait for the French government to declare muscular dystrophy a national priority. Researchers entered into dialogue directly with patients – and on their own initiative. Nor did they feel it was necessary to institutionalize the process by establishing a university department of muscular dystrophy before work could begin. Both patients and researchers behaved experimentally, exploring together the concrete implications of what came to be a shared vision of the possibilities. Both may have believed they were placing their trust in Mode-1 science, but in their interaction they produced a Mode-2 environment.

A transaction space was created in which both patients and researchers had something to exchange. Each could bring something to, and take something away from, this space. There were powerful constraints, perhaps the most threatening of which was the lack of obvious funding. There were no earmarked resources, as there would have been in the case of a national research programme. The creative alternative, raising money for research through a national telethon, worked – but success was not guaranteed (as it would have been with a national programme). Success, or failure, depended on the context. If the telethon had failed, the initiative might have floundered and the transaction space would have closed. Society, in the form of muscular dystrophy sufferers, was able to speak to science, in the shape of interested bio-sciences researchers. A new focus for biochemical research was generated. And the dialogue continues. Further telethons have been held. The research focus has now shifted from biochemistry to genetic studies using molecular biological techniques.

When comparing this case with other cases described in the literature – like research into retinitis pigmentosa (which is one of the main causes of blindness in industrialized countries and is still incurable) we see similar patterns emerge. But there also appears to be a variety of forms of co-operation between patients and researchers; ranging from 'patient-dominated' forms to 'researcher-dominated' ones. The open question in both types is, of course, how the co-operation develops over time. If, at the time of its foundation, the patients' organization is confronted with an existing specialized scientific or medical society, it tends to rely on these research activities which are also often more resistant to open up to and 'translate' the patients' demands and concerns into their research. If, on the other hand, patients' organizations become sufficiently self-conscious to exert their influence, the chances of strongly contextualized knowledge developing are far greater (von Gizycki 1987). In the next chapter we shall discuss how such a transition is occurring.

These examples – the Central Artery/Tunnel Project in Boston, and co-operation between medical research and patients' self-help groups – display some of the characteristics of strong contextualization outlined in this chapter. First, they appear messy judged by the conventional rules either of how large-scale technological projects are planned or research is normally conducted. No single person or agency is in charge and management is exercised in a more inclusive, participatory way. It is these qualities and characteristics which allow the potential of these projects to be realized under existing constraints. Second, participation is not a single event, a one-off affair or a preliminary ground-clearing exercise. Instead it is iterative – views are solicited, advice sought, designs modified and then the whole cycle is repeated. Nor is the involvement of researchers in a relatively rare disease, which does not warrant sufficient investment of the pharmaceutical industry to engage in long-term and costly research, given once and for all. It is on trial, as is the patients' involvement. Third, the outcome – whether new approaches to the (often non-existent) treatment of such diseases or the building of a new highway in Boston – is judged by both experts and society, or researchers and patients, to be better (in scientific and social terms) than the results that would have been achieved by following more traditional methods.

10

Contextualization of the Middle Range

We have argued that contextualization is a consequence of at least three conditions, which may operate at different levels and not all of which have to be present at the same time: the overall shift (or drift) from a model of 'segregation' to one of 'integration'; the selective retention of certain potentials which arise as a result of greater variation; and the place accorded to 'people' in our knowledge, be it as actively involved in its production or conceptualized as either objects of research and/or as addressees of ensuing policies. In this chapter we will argue that the process of contextualization which transforms potentials into concrete research activities often takes place in what we call transaction spaces (generalizing from Peter Galison's concept of the 'trading zone'), and that these can be distinguished by the intensity of communication between the various actors that occupy them.

In terms of the evolutionary perspective we have put forward, weak contextualization is characterized by communications patterns that are determined largely by institutions or representatives of institutions. In other words, 'people' are aggregated and their wishes and desires are in a sense represented by institutions, whether the state, industry or science. They are unlikely to play a major role, nor does 'society' enter much in other ways. By contrast, strong contextualization is characterized by an intense involvement even in the planning phase with those who are also most likely to be affected by the research, and the hope that this will lead to a better outcome. Taking an example from the engineering and management of technological mega-projects is helpful to illustrate how far contextualization can go in 'making a difference' not only in organization and management, but in the engineering design of the project itself. Another candidate for the future is medical research which closely involves patients, as well as the field of environmental research – a lively, sprawling Mode-2 research activity, if ever there was one.

But the process of contextualization is mainly to be found in a middle range between weak and strong. This is partly due to the great variety and heterogeneity of research fields and actual research practices, but it also reflects the fact that the 'how' of contextualization may be better observed by not trying to force it into a single mould or ideal type. There are two preconditions that clearly favour the middle range. The first is the emergence of 'transaction spaces' between different groups, disciplines, research fields or other major configurations. In these spaces transitory transactions take place in which supposedly all parties have something to gain – or an interest in discontinuing the process. Intrinsically linked to these exchanges is the potential emergence of Mode-2 objects. These are devices or other research objectives in the making that crystallize the transaction process and help to sustain dialogue and negotiation. Of course, Mode-2 objects are not present in every instance, but their appearance can greatly facilitate contextualization. The presence of Mode-2 objects also makes it easier to identify when contextualization of the middle range is taking place. Accordingly, we provide examples of where Mode-2 objects have emerged, and where they have not. Lest our presentation give the impression that contexts always exist prior to contextualization, we add an example of how a previously non-existing context was mobilized while contextualization was simultaneously taking place.

The second precondition for contextualization of the middle range has to do with moving beyond the context of application and setting one's anticipatory vision on what we call the context of implications. This extension of the research process beyond what is traditionally considered to fall within the legitimate concern of researchers marks an important and novel departure in the direction of contextualization of knowledge, limited as it still may be at present.

From Galison's 'Trading Zone' to Transaction Spaces

The rise of hybrid *fora* was discussed in *The New Production of Knowledge*. These hybrid *fora* are public spaces where, for example, the risks associated with certain technological developments are debated. We claimed that in these *fora* important new knowledge was being produced. But other spaces can also be identified, in which encounters among various kinds of actors, with different interests and outlooks, have become possible and which carry implications for the process of knowledge production. It is now recognized that interesting and challenging science can be produced outside disciplinary structures, giving rise also to changes in curricula and hence in the transmission of scientific knowledge. Instrument technology has also become increasingly significant for experimental design, simulation and data collection and analysis. The pervasiveness of such instrumentation and techniques across a vast range of research fields, specialities and sub-specialities is opening up new channels for the transmission of concepts, ideas and results which, in turn, generate further techniques and procedures. This is what underpins the dynamics of science and leads to further acceleration of the development of new scientific knowledge.

In Mode-2 society new spaces are opening up where, because of intensified competition, greater experimentation with potential partners in knowledge production is under way. Entrepreneurial attitudes, including the crucial entrepreneurship needed to acquire the resources necessary to carry out research, are rapidly developed. Such attitudes, as we have shown, may help to create conditions in which reflexive interaction between science and its context is stimulated. These new spaces provide an important framework in which still tentative, and as yet inadequately institutionalized, interactions can take place. But these interactions are more than random encounters. They can develop into genuine transaction spaces which strongly recall some of the essential features that Peter Galison has described for the 'trading zones' he came across when analysing the history of nuclear physics in the twentieth century (Galison 1997). In this magisterial work we are made to encounter the fascinating exchanges and intense collaborations between three sub-cultures of physicists – theoreticians, experimentalists and engineers (who built the machines used in nuclear physics). These traditions remained intact, preserved inside the collaboration, while the co-ordination of exchange took place around the production of the two competing instrument cultures of 'image' and 'logic' which ultimately joined. Taking his lead from anthropology, Galison observes how their often asynchronous exchanges can be compared to the incomplete and partial relations which are established when different tribes come together for trading purposes. Nothing in the notion of trade presupposes some universal notion of a neutral currency. Quite the opposite: much of the interest of the category of trade is that things can be co-ordinated (what goes with what and for what purpose) without reference to some external gauge. Each tribe may bring to this interaction and take away completely different objects as well as the meanings attached to them. An object which may have a highly symbolic or even sacred value for one tribe may represent an entirely banal or utilitarian object for another. Nevertheless, interaction and trade is possible and actually takes place – to the obvious benefit for all because, if this were not so, dialogue would have ceased.

Trading may also give rise to the emergence of contact languages, like 'pidgin', as a means of communication which is inevitably incomplete and truncated. Likewise, physicists and engineers were not engaging in translation as they pieced together their microwave circuits, nor were they producing 'neutral' observation sentences. They were working out a powerful, locally understood language to co-ordinate their actions. Despite obvious limitations, some kind of understanding and exchange does occur in such situations. By borrowing from anthropologists the metaphors of 'trading zone' and of the rise of 'pidgin' language (which might, under certain conditions, become a proper language, like Creole), Galison incidentally calls into question part of Thomas Kuhn's thesis about paradigm change, because he shows that in Kuhn's version, paradigm change was essentially limited to theoreticians. In reality, however, nothing prevented experimentalists from continuing their interaction with theoreticians and vice versa, and the same was true of machine designers and the other two sub-cultures. Incomplete and partial, as well as transitory, as these exchanges might all be, the history of physics and of its accomplishments shows that communication is possible (laying Kuhn's incommensurability thesis to rest) and that the establishment of 'trading zones' plays an essential part in co-ordinating (sub-cultural) belief systems and action (Galison 1997).

We want to extend and generalize the metaphor of the trading zone beyond interactions among scientific sub-cultures to wider exchanges that take place across disciplinary and institutional boundaries. These we regard as transaction spaces. The idea of 'transaction' implies, first, that all partners bring something that can be exchanged or negotiated and, second, that they also have the resources (scientific as well as material) to be able to take something from other participants. Of course, the meanings attributed to exchanged objects may greatly differ for different participants. But the success of these exchanges depends on each participant bringing something that is considered valuable by someone else – whatever that value might be. Participants usually will return to their 'home-base' with their gains, thereby reinforcing in typical Mode-2 fashion the links and exchanges that have already occurred by sharing with others. This notion of transaction spaces makes the evolutionary process which was discussed earlier much more specific, because these transaction spaces are where the first tenuous interactions between 'society' and 'science' take place. They are spaces (both symbolically and very concretely) where potential participants can decide what might be traded and also establish the lines of communication necessary to sustain discussion of potential to the point where constraints become visible. Of course, if the constraints are too severe the transaction space may disintegrate. But through further interaction, ways may be found to overcome constraints – and a more robust research activity may emerge. So the growth of transaction spaces, some of which persist while others are transitory and temporary, is characteristic of Mode-2 society's interaction with Mode-2 knowledge production.

Mode-2 Objects

One of the outcomes associated with contextualized knowledge and the transaction spaces in which it takes shape may be the emergence of Mode-2 objects. Even if they are adumbrated in the beginning of an exchange or transaction process, Mode-2 objects are hard to specify or predict because they may emerge only towards the end of the process. The process is more one of groping towards this 'object of negotiation', which has yet to assume its scientific or technological *Gestalt*, than of knowing from the beginning what its contours and content are likely to be. In the case described by Galison one of the most important Mode-2 objects to emerge from the trading zone which engaged the three scientific sub-cultures of theoretical physicists, experimentalists and the machine builders turned out to be the particle detector, a key device underlying the achievements of nuclear physics. While the emergence of Mode-2 objects does not explain how contextualized knowledge comes about – and they certainly are not a necessary outcome of such multi-dimensional transactions – they may nevertheless be one of the most important outcomes of the process of contextualization. In retrospect, they may look like the entities which were necessary to make things 'click'. Yet, the converse is also true: when things click, Mode-2 objects may emerge. Thus, Mode-2 objects may be a good indicator that some form of contextualization of the middle range is occurring.

The Human Genome Mapping Project

An example of the emergence of such a Mode-2 object can be seen in the UK Human Genome Mapping Project (HGMP) as analysed by Balmer. It should be recalled that the aim of the Human Genome Project was to draw up a catalogue of the entire genetic make-up of the human genome. The maps, like geographical maps, could be of varying type and resolution, from large-scale linkage maps of genes in relation to other genes based on frequency of co-inheritance, through various types of physical maps that locate 'landmarks' in the DNA, and eventually to the highest resolution, the sequence of chemical base-pairs which make up the DNA molecule. The project did not come about without controversy. Proponents of the project claimed that it would provide a valuable resource for science and medicine, while opponents challenged its wisdom in terms of cost, strategy, ethics and the ultimate utility of its results (Balmer 1996).

Balmer examines the mutual shaping of science and policy, arguing that the evolution of the policy was not merely a series of administrative choices. The strategies, boundaries and accounting methods of the programme, together with the organization and conception of research among grant-holders, reflect the often competing demands from the world of science and administration. Both the style of research favoured and the degree to which grant-holders were held to the project mission directed research in a more subtle, contextual manner. The HGMP was not the outcome of any single factor, nor does it follow the model of national programmes of the type we have outlined in the previous chapter. The fact that a mapping project emerged has to be understood not in terms of bureaucratic politics, but as the outcome of a complex process of negotiation in which a large number of interested parties were involved. While some actors and institutions played a large part, others were less important. The point is that no single person, group or organization was in control dictating the pace and direction of advance. Rather, the UK policies of 'selectivity and concentration' and 'value-for-money' provided the guidelines for a co-ordination with the agendas of the Medical Research Council (MRC) and the gene mapping community and their spokespeople. As Balmer describes the process, the Human Genome Mapping Project came to act as a Mode-2 object (a boundary object, according to Balmer).

The project constituted a social and a political entity that was able to align the goals and agendas of separate working groups. Alignment was achieved over a period of time as groups and their agendas were shuffled into and out of the policy arena, or altogether marginalized. As a consequence, money flowed from the state to scientists, and gene mapping under the auspices of a concerted organized programme came to be supported. As the programme developed, what was at stake was the continual need to maintain credibility – to be seen to be funding good (or worthwhile) science while at the same time implementing a directed, mission-oriented programme of research. The question of what science to support was translated into debates over the best way to do science, what scientific knowledge was for and what was to count as worthwhile knowledge.

The emergence of a Mode-2 object can be crucial in the contextualization of knowledge in the middle range. Something is necessary to align diverse and often divergent interests if work is to get started, but it is not a planned process. At each stage of the development of a project, the contingency of events and the opportunism of the actors could not be ignored. Scientists (Brenner and Bodmer, in this case) may have had some degree of control; the government, together with the civil servants of the Advisory Council for Science and Technology (ACOST), and Medical Research Council administrators may also have had their say. In sum, it was more of an orchestration process that attempted to make use of the resources available than a planned strategic, networking process which led to the emergence of the HGMP in the United Kingdom. In this case, Balmer concludes, the process was like an orchestra in which all the players vied to be the conductor, but with no one fully in control and everyone ready to improvise. The Mode-2 object – the genome mapping project – allowed some sort of melody to be heard. (There have probably been no equivalent 'objects' in Mode-1 science – or if there have been, only as abstract, theoretical or mathematical entities.)

In this case, one can observe how the social, economic and scientific strands were woven into a mapping project and how important it was to have something to command the allegiance of diverse interests in order for the project to be carried forward. A complex set of negotiations took place over an extended period of time, and these were guided by a particular form of organization which acted as a vehicle around which many diverse interests could converge, at least long enough to allow funding to flow. Such contextualization may appear to be largely a spontaneous or even a self-organizing process (and the influence of the more global development of genomics projects on the UK effort cannot be ignored either), but, while it is true that many individuals from different backgrounds and with different perceptions saw the benefit of the mapping project and agreed from many points of view that it was desirable, it was greatly 'facilitated' by a Mode-2 object. As the subsequent events show, the project eventually ran into problems generated, in part, by the emergence of other 'competing' Mode-2 objects which promised quicker scientific results more cheaply (Carr 1998).

The case of hypersonic aircraft: The absence of a Mode-2 object

The absence of a Mode-2 object presents different difficulties. For example, artificial intelligence never acquired such an 'object' which could orchestrate its various strands, which even today remain divergent. Informatics, as a result, has never cohered. Another example, that of the development of a hypersonic aircraft, provides an illustration of what can happen if no Mode-2 object emerges. In the case of hypersonic transport we can note a general evolution of aircraft development in terms of the top speed of the vehicle – sub-sonic, supersonic and hypersonic. In each case the characteristics of the vehicle must reflect the properties of the flight medium at that particular speed: a sub-sonic vehicle will not perform well at supersonic speeds and vice versa. The social and institutional framework for the development of each type of aircraft is different; a different range of scientific and technical experts, a different set of military and industrial interests, regulators, noise abatement protesters and different configurations will dominate in each case.

Across these frameworks, there is a general agreement that a hypersonic aircraft is desirable and that it should have both military and civilian uses. The problem is how to get started. It is acknowledged that the relevant science is not in place. What is more, the science cannot begin until certain technological explorations are carried out that will structure the scientific domain so that appropriate flight equations can be derived. But the structuring of that domain cannot proceed without a whole series of previous judgements as to both the civilian and material uses envisaged. Use will, of course, dictate the shape of a desirable commercial or military vehicle. But, to complete the circle, the range of possibilities is, in turn, conditioned by the findings of scientific research which are themselves related to the development of new instruments. In other words, a technological infrastructure is needed. We can see in this example both the sources of the uncertainty as to how to begin as well as the interdependence of sub-cultures of theory, design and commercial/military considerations, each of which operates in a particular bureaucratic framework. It is expected that out of their mutual interaction will come path-breaking science, but the question is how to get started?

In this case, the context might be regarded as being dominated by the economic power of military or commercial interests, by their perceived need of the desirability of having a hypersonic aircraft. This can be true only in a very general sense, because at present there is little to guide these participants as to the rate or direction of activity, and without more information it will not be possible to invest the vast sums required. It is not so much a question of finding resources as of deploying in a reasonable manner what already exists. Equally, without some investment the relevant scientists, technologists and designers will lose interest and direct their energies elsewhere. In a sense, and at the time of writing, the contextualization of knowledge has not gone very far. What is lacking seems to be an appropriate Mode-2 object which might align the various interests rather in the way the Human Genome Mapping Project did in our earlier example. Not every context, therefore, finds itself hosting the production of knowledge in an equally accommodating way.

Design configurations: Mode-2 objects in industry

In developed economies, society has always spoken back to firms in many ways, but perhaps primarily using market signals. Of course, markets respond to 'people' at different levels of aggregation: to individuals when they choose consumer goods; to companies when they order capital goods; or to governments when in their representative role, they impose regulatory regimes. Through manifold, multi-level interactions, markets respond with flows of products, whether they be goods or services. In so far as these products depend on knowledge, and the current wisdom of the new theories of economic growth is that the competitive advantage increasingly lies precisely in a firm's ability to acquire and configure knowledge to its ends, that knowledge would be contextualized knowledge (Dosi 1996). It is perhaps, then, uncontentious to say that industrial R&D, because it must communicate with society through markets, yields contextualized knowledge. Markets typically operate on final products, and market signals tell firms not only how many of each product they can sell at what price but also whether the product has the appropriate range of performance characteristics. In the latter case, markets inform firms that their products need to be improved, and it is in searching for ways to achieve these improvements that industrial R&D can be said to produce contextualized knowledge.

Still, society comprises more than markets, and society makes itself heard not only through the ebb and flow of quantities and the fluctuations in the prices of goods and services but also through the constraints of a more comprehensive selection environment. For each firm, the selection environment comprises, in addition to its main competitors, conventional market institutions and sources of finance, other regulatory and environmental bodies and, increasingly, pressure groups with views on the quality, reliability and safety of its products and processes. The imperatives of a selection environment involve a different type of signals than those conventionally associated with the 'market'. So, for example, contemporary firms need to take on board not only changes in national policies concerning currency movements, interest rates or other fiscal legislation but also an increasingly wide range of social factors, including environmental and health and safety issues and employment protection; and increasingly they must become familiar with the ethical imperatives of different cultures.

In legislative terms, many of these constraints are the outcome of government policy but they may also arise, more or less directly, from the activities of consumers and other special-interest groups who usually have no formal relationship with the government. For example, the decision by Shell to dispose of a disused oil rig by dropping it onto the sea bed was reversed not by government legislation but in large part as a result of the pressure brought on the company by the environmental pressure group, Greenpeace. In a sense, then, the selection environment is the source of multiple communications between society and the firm. This produces a set of constraints with which in making technological choices each firm will ultimately have to come to terms. In particular, these constraints can shape the search for knowledge, prompting various experimental approaches that aim to get round or overcome them. In so far as each firm produces knowledge germane to its own situation it can be said to produce contextualized knowledge. How does this process work?

That firms have a constant need for specialized knowledge cannot be controverted. There is no product, process or service that does not need to be improved – competition itself guarantees that – and these improvements will be the result of the knowledge derived from a better understanding of the constraints of the selection environment or of the potential of technologies that it is using, or perhaps both. Still, much of this knowledge is generated in-house. In this respect, all technological innovation rests upon the production of knowledge, though it may remain more tacit than codified knowledge. But there is another, deeper process at work in each firm. For most firms, its nest of core technologies cannot be bought 'off the shelf' from someone else. They need to be identified, then developed inside the firm. The choice of core technologies is the result of a decision-making process in which a firm chooses a particular design configuration and directs its resources to exploiting its possibilities. Such a choice, unless the firm is very strong financially, usually means that other possible design configurations will not be followed up.

A design configuration, though it may look like a technical object, is in fact a socio-technical device. It sets up both an organizational and a technological arrangement within which the firm will both identify and develop its competences. It sets out the range of materials and specifies the knowledge and skills that the firm needs to acquire. It not only provides a focus for the ongoing development of business strategy but also provides initial heuristic guidance as to the likely impact of this choice on various aspects of its selection environment. The design configuration is the device from which, in future, a stream of products or processes as well as their improvement will emerge and through which ever shifting imperatives of the selection environment will need to be incorporated. For each firm, the choice of a particular design configuration is crucial not least because it is aware that its principal competitors are engaged in precisely the same process and each realizes that its survival depends upon making the correct choice. At a deeper level than is conventionally implied by the term in economics, competition between firms is competition in the search for robust design configurations.

Further, the choice of a design configuration is, in practice, also the choice of the team that will develop it. Increasingly, it is recognized that much, perhaps most, of the requisite knowledge and skills lies outside the firm and for that reason the development of a design configuration is a matter of becoming involved with the right team. At this level, competition between firms is essentially a matter of competition between different teams each of which is engaged in knowledge production. When it emerges (if it emerges), it is the design configuration that will give the orientation to the firm's future knowledge-producing activities. It will determine the range of expertise needed and suggest the most likely places where it can be found. It also suggests the sorts of skills which must be developed in-house.

In a sense, a design configuration is a Mode-2 object. As with other Mode-2 objects, the design configuration has the ability to 'align' the efforts of different specialists, often from different institutions, until such time as either the design configuration is found or a better option emerges. In the latter case, the unsuccessful teams may dissolve or enter into other problem-solving contexts. Utterback has shown how selection environments and design configurations can interact and lead, in some cases, to a dominant design which a number of firms seek to exploit, though each may do so in a slightly different way (Utterback 1994). But this refinement ought not to obscure the central role played by the identification of Mode-2 objects in producing contextualized knowledge. As with other Mode-2 objects, design configurations do not already exist, they need to be 'discovered', and this process follows the general lines of the evolutionary perspective that we have already described.

With this in mind, the 'how' of the contextualization process – at least as regards the behaviour of firms in a particular industry – may be clearer. It is the firm's knowledge base, including those additional human resources which it is able to access, that identifies those potentials and explores whether or not they can be accommodated to its selection environment. The result may be the emergence of a Mode-2 object, a design configuration of sufficient coherence to allow further collaborative work to test its robustness. Clearly, a design configuration will be useful only to the extent that the context – the key elements of the selection environment – has been incorporated into it. Failure can come from either side. To ignore or fail to take sufficient account of a particular aspect of the selection environment can be just as damaging as misjudging the development potential of a particular piece of science or technology. In this somewhat narrow but none the less very practical sense, then, the only reliable knowledge is contextualized knowledge. In terms of the language we have been using, design configurations are Mode-2 objects; they are the outcome of an intense experimental search process in which certain potentials are articulated under constraints of a specific selection environment.

It is through the selection environment, then, that society speaks back, and it is communicating a more complex message than is usually implied in the notion of markets as sites for commercial transactions. In searching for a robust design configuration, knowledge is produced from a variety of individuals drawn from a range of institutions. It could, therefore, be argued that the knowledge so produced has some of the characteristics that we have identified with strong contextualization. However, because society speaks back essentially through selection environments that are still dominated by institutions where 'people' enter the research activity only at highly aggregated levels or where their wishes are generalized through market research – for example in relation to health and safety issues – such communications are more characteristic of weak contextualization. It is for this reason that we place this type of Mode-2 object in the category of contextualization in the middle range.

The Emergence of a New Context: The Case of High Temperature Superconductivity

So far, we have concentrated on the emergence, or non-emergence, of Mode-2 objects, while the background assumption had been that a context existed and was ready to exert some selective constraints on new possibilities. But in a world where science and society move increasingly towards more 'integration', the context also may still need to take shape before, or at the same time as, contextualization. The following example illustrates how this can happen.

The discovery of high temperature superconductivity (HTS) is an instance of how a context which did not exist prior to the event might emerge subsequently. While we hesitate to see in HTS a Mode-2 object, its discovery – made in an IBM laboratory in Switzerland – swept the world of science along on a wave of growing expectations regarding its technological potential. One may surmise that the discovery of a phenomenon that for a long time had been believed to be unattainable at high temperatures served to mobilize many different scientific and technological as well as economic or medical strands, in pursuit of these technological possibilities. Focusing on the immediate aftermath of the unexpected scientific breakthrough, Nowotny and Felt (Nowotny and Felt 1997) analysed the emergence, after the discovery of HTS, of a new research field which cuts across existing disciplinary boundaries, creating new forms of scientific co-operation and organization, while mobilizing funding wherever it could be found. Following the discovery and the enthusiastic belief in its technological potential, a broad and unintended 'collusion' occurred that aligned researchers with different disciplinary backgrounds and experience (in the 'old', low temperature superconductivity field) with decision-makers in funding agencies and ministries, politicians who were quick to seize the opportunity for their own ends, the mass media that had found a good 'story' and, last but not least, the general public. The overall effect was to promise more than could reasonably be delivered in the short term, but it was also clear that a context was in the making.

The case of HTS is interesting, first, because it provides an instance of what can happen after an unexpected discovery has been made. For many scientists, discovery is still the most important thing, with subsequent developments being regarded as applied science. In the case of HTS, the line between basic research and potential applications ceased to exist right from the beginning. HTS can be seen as a case that shows how complex and fluid the present situation in the transformation of the science system has become. Researchers can no longer expect to find an environment hospitable to their work, but are compelled to create one. It takes extraordinary effort, time and energy to set up the conditions under which research programmes can run for a predictable period. Such efforts are no longer external to, but have become an integral feature of, scientists' work. The situation is also extremely fluid on the level of policy-making. Nowhere is science policy firmly in place; rather it is continuously shaped and reshaped under the impact of old and new requirements, such as priority setting, selectivity, strategic thinking in research and future exploitation of scientific knowledge produced. Basic science, as the case of HTS clearly shows, has inexorably been drawn into the dual orbit that marks science at the turn of the twenty-first century: basic science has become useful and it (therefore) is expected to enhance international economic competitiveness. This is another level where we can see how contextualization of the middle range proceeds – relentlessly.

Second, if there was a general belief amongst those involved with the original discovery that applications would quickly follow, it was by no means clear how these expectations would be realized. This opened up exciting opportunities for many who believed that they were called; but as it turned out, only few would be chosen in the end. But in the meantime, HTS and its potential for technological innovation had to be, and was, successfully 'sold' to the public – by politicians who themselves had been sold the prospects of proudly riding the wave of the next key technology. It was sold to and by the media as bearing the potential for levitated trains, energy-saving power transmission systems and entirely new gadgets. Dressed up in the garb of its future technological benefits, HTS had the advantage of convincing any lay person why, in this case, funds for basic research were needed and deserved. Other participants colluding in buying and selling technological promises were the research councils and government ministries responsible for funding research they consider vital for the national interest in a globalizing economy. Scientists therefore were not the only 'salesmen'. Neither were they the magicians the media sometimes portrayed, nor the powerful lobbyists that are associated with 'big science' or that emerge in the course of prolonged mutual service with governments.

The point is that no one can be singled out as originally 'selling' HTS basic research for its potential, but everyone took part. Ludwig Wittgenstein might have termed what took place a 'truth game', in which each participant has his or her version of an elusive but shared belief. It was obvious to all who took part in the game that what they hoped to engage in, fund or set up and manage would remain in the realm of basic science for years to come. No serious researcher or science administrator, no journalist adhering to professional standards, could or did claim otherwise. And yet every public presentation and representation of the newly emerging research field, every speech or interview, every single research proposal in every country, every international conference resonated with the hopes and expectations that technological benefits would come, sooner or later. The time-horizons expected for the different types of technologies were discussed publicly and privately, and various estimates were published. This truth game was surprisingly high in truth content, but it remained a game, a collective bet on an uncertain outcome. What gave rise to the fervent, pervasively shared enthusiasm and belief? There was no proof to substantiate the belief, and no one claimed that it could be substantiated. It was another instance of contextualization of the middle range – this time, observed with the context-in-the-making.

Third, HTS allows us to see the innovation machinery of 'science and society' at work – in a context of application where neither the context nor the applications are as yet in place. HTS has revealed the close interlinkage between basic science and technology, and how the anticipation of technological advance affects the choice of what basic science will be pursued. The 'extended lab' – the vast and heterogeneous network linking each local laboratory to its economic, political, administrative, technological and scientific environment, where each network includes many partners who shape and define the content of research, orientation of programmes, and evaluation of results (Callon 1989: 13) – continues to expand, neither according to a master plan nor by being left to chance.

The example of HTS puts a different gloss on the contextualization of knowledge by showing how a context which did not pre-exist can form after a major event – the discovery of high temperature superconductivity, in this case. There is little doubt that the discovery itself emerged in what we call the context of application. Indeed, there was a long history of those who had forecast a major fallout of applications long before the discovery was made. This is not an unusual occurrence in science, but here we have an illustration of how a previously non-existent context emerged within a very short period of time once the unexpected discovery was made and evolved. In a sense, following the discovery a new context had to be generated to allow further developments to take place. The evidence shows that this was not the result of strategic planning but took place through the enthusiasm and conflict that characterized the evolution of a framework in which many actors – from inside and outside science – could, and in fact had to, participate in order to align various interests and to set up a collaborative-competitive regime where inputs were diverse and outputs as yet unknown. It also shows how important shared beliefs – even an unintended collusion – are in underpinning such efforts. They were like glue, holding together the otherwise 'seamless web' of scientific knowledge and curiosity, of technological potential and the social fabric of competition and co-operation that came together in making up the new context. It is important to recognize that neither the scientific nor the human or organizational resources existed 'ready to hand' at any given time. These resources were acquired, configured and re-configured on a continuing basis. A context-in-the-making cannot take anything for granted; it must be generated by the participants and be capable of maintaining itself.

It is increasingly recognized that creativity and innovation depend on a self-organizing capacity. In the search for technological innovation, specific products are no longer targeted; instead infrastructures and research conditions are planned with the intent to allow certain types of products to emerge, including those whose uses are as yet unknown. But what conditions favour scientific and social creativity and innovation, given the unpredictability of the research process and the complexities of social life? What does HTS show about the collective research response to the unpredictable? How well prepared are funding agencies, research councils and similar bodies when the unpredictable breakthrough occurs? How well prepared are research groups to exploit new opportunities? How well prepared are countries to redirect their organized research efforts?

Contextualization of the middle range touches closely upon all these questions. The one generalization safe to make is that preparedness depends on the strengths and weaknesses of the historically evolved structures in which research is conducted. It may involve the research system's flexibility, including the components' flexibility toward each other. The degree of preparedness to grasp new opportunities varies markedly between research groups, industries and countries. At the end of the day, it turned out that preparedness was crucial. The research groups that survived and, after much turmoil, were able to ascertain their scientific standing, were those who were prepared and able to use this resource to its maximum extent.

Exploring the Context of Implication

Nothing could perhaps better characterize the shift from Mode-1 to Mode-2 research than asking what happened to the famous distinction introduced by Popper (Popper 1969) between the 'context of discovery' and the 'context of justification' which was used as a convenient epistemological shorthand to summarize how Mode-1 research was perceived to function. In contrast, we have emphasized the importance of a 'context of application', in which knowledge production in Mode 2 takes place. Yet, stimulated and reinforced through more frequent interaction with a Mode-2 society, research activities now transcend the context of application. They begin to reach out, anticipate and reflexively engage with what we call the 'context of implication' – those further entanglements – consequences and impacts – that research activities continue to generate. The context of implication thus always transcends the immediate context of application in which it occurs. It may reach out to neighbouring research fields and to as yet obscurely recognized future uses. While nobody can know precisely when and where a particular implication will arise or what will result from it, and while it is virtually impossible to assess its importance, there can (and perhaps ought to) be a forward look, a serious attempt to reflect and anticipate what the context of implication may hold – however much uncertainty may enshroud the effort.

Taking the context of implication seriously opens the door to people – to their perspectives, on the one hand, and to their constitutive role in research activity, on the other hand. In the first instance, people may be encountered haphazardly as individuals, perhaps as colleagues or rivals. They may come from other scientific disciplines or materialize more concretely than the otherwise hazy category of 'users' suggests. But one may move from such random involvements and ask who ought to be interested and/or implicated in a research activity – and start to act accordingly. Whatever results from such an exploration of the context of implication and however uncertain the insights that may be gained – for ultimately only history will tell and there will always be more than one story to be told – such inquiries open the door to people, allowing them into the space in which knowledge is produced. For ultimately, scientific knowledge will need to be tested not only against nature, but against (and hopefully also with) people.

Tracking the Human Genome Project

The United States Congress requires that the US Human Genome Project devote 3–5 per cent of its annual budget to social impact studies which are meant to draw out social, legal and ethical implications of ongoing work on the human genome. In 'Tracking the

Human Genome Project', a group of three anthropologists set out to map new genetic knowledge among three constituencies: research scientists, clinical physicians and patients living with the diseases and disorders that have become the objects of genomic investigation. By engaging the 'natives' – in this case the scientists – at the centre of their belief systems and practical activities, by following them in their laboratories, at collaborative meetings, on the Web and in corridor talk, but equally by watching and listening as clinicians diagnose and treat rare conditions that run in families and by attending meetings of support groups and voluntary organizations, they also learned how knowledge that patients acquired from actually living with a genetic condition is put into social practice. The field of social practice turns out to contain an enormous range of individual and social difference, even though it originates with a small set of genetic alterations that can be grouped into a neat set of scientific problems.

Out of the vast research space which the genome constitutes, the project tracks only three connective tissue disorders: chondrodysplasia, or dwarfing condition; Marfan's syndrome, or ocular and skeletal abnormalities; and EB, epidermolysis bullosa, a family of blistering skin diseases. What constitutes a 'genetic success story', however, varies dramatically with the condition, the degree and kind of medical consideration it receives, and the social circulation of information, aspiration and practical knowledge about it. Eugenic fears, for instance, are most clearly expressed among those afflicted with dwarfing syndrome, while patients with Marfan's syndrome tend to be extremely receptive to genetic diagnosis, the circulation of genetic information and aggressive medical interventions. Moreover, the stories of these different groups of patients have to be situated in a much larger framework that includes attention to kinship, work and community relations. Trying to understand the context of implication makes one acutely aware of how a particular disease transforms the way people think of themselves and also about those who are participating in the genomic exercise, about who is positioned truly to conduct research or give informed consent, about who benefits and who is burdened by new genetic knowledge, and the extent to which much older forms of social differentiation and stratification survive (Rapp, Heath and Taussig 1998).

The picture of the contextualization of knowledge that emerges here is one in which researchers are increasingly becoming engaged – directly through their own awareness and indirectly through the numerous mediating organizations and patients' groups that spring up – with the context of implication that is shaped by their research, while simultaneously shaping the kind of research that will be undertaken in the future. With the benefit of hindsight, in this case provided by social scientists, an ongoing middle-range contextualization is observed, described and analysed. (The main difference from the kind of interactions between medical researchers and patients' organizations, as described at the end of the last chapter, lies in the emphasis which here is directed towards learning more about patients' reactions, fears and hopes, and the explicit willingness to change research directions in order to make room for this kind of input. The context of implications thus can be said to shape also the content of research in a more or less decisive and ongoing way.) Whenever it is possible to 'feed back' the insights thus gained to those who were involved in the first place, the context of implication extends backwards and leaves its traces in the research undertaken. By making the different groups (genetic researchers, clinicians, patients and their families) aware of each other's different perceptions, interests, outlooks, hopes and fears, otherwise single-minded views can combine into a truly pluralistic perspective, which may allow people to be given a place in the production of contextualized knowledge where originally no such space was foreseen for them.

The writing of history: A tension-ridden dialogue with the context of implication

History and historiography have never been straightforward affairs. Ever since Thucydides, it has been remarked that it is usually, but not always, the defeated who feel an urge to narrate and to account for what has happened to them. The questions that historians find interesting, as well as what are considered to be legitimate topics for historical research, have therefore always been closely intertwined with the wider societal context. Since the lessons to be drawn from history are few (and are more often than not highly ambiguous), it is not so much the context of application, but the context of implication which matters most to historians. Historians are engaged in a continuous, often tension-ridden dialogue with the surrounding society. Their research agendas result from intricate and at times also highly controversial and emotional exchanges, while historical research undoubtedly contributes to a country's or a social group's sense of identity. Nowhere is this more in evidence than in those instances where the research questions, because they 'go under the skin', have deeply troubling implications for the rest of society.

The two examples, briefly presented here, illustrate how middle-range contextualization of knowledge occurs in the social sciences. The first example comes from a new awareness that the writing of

African history may not need to follow the hitherto dominant model of European history. Under its strong dominance, the past of other regions of the world, or even of a continent, that would not fit European ideas, European techniques or European examples was disregarded or marginalized. In the case of Africa, given its relative paucity of written sources its history seemed to have little to offer for the understanding of 'How have we come to be where we are?' A younger generation of African historians has taken up the challenge, answering the question with an armoury of new methods and techniques. In order to circumvent the limitations of their archival material, they had to break with the usual conventions of European and North American historiography. The new African historiography had to draw upon a much wider range of materials, beginning with testimony transmitted orally from one generation to another, carefully studying its narrative and poetic conventions and, whenever possible, checking and comparing it with external written sources, or other, archaeological or linguistic, evidence.

In a next step, Africanists went beyond comparing written and oral evidence. They brought together techniques from archaeology, anthropology, botany, chemistry, ecology, economics, genetics, linguistics and sociology in order to widen the boundaries of historical understanding of Africa's past. Dendrochronology and carbon-dating are by now commonplace techniques. Historical linguistics in combination with modern genetics were used to tease out the broad history of African migrations. Palaeo-botanists looked for patterns of change in climate and ecological conditions, so important for shifts in agriculture, demography and the rise and fall of human settlements, and allowing the reconstruction of other aspects of social, cultural and political life. Thus, a new approach to history emerged, making much fuller use of other disciplines and research techniques. But it came in response to the relative neglect of research questions which were not considered to fit the dominant model of how history should be written (Appiah 1998).

In this example, a latent context of implication existed, but it had to become mobilized. This was achieved through a general shift in awareness and a successful attempt at breaking out of the European-dominated model of historiography. The challenge proved sufficiently strong to induce a younger generation of Africanists to extend, rather than retreat, into new scientific territory, thereby greatly expanding the scope of conventional methodologies and the basis of what is considered relevant historical evidence.

The second example deals with recent changes in the kind of questions asked about the history of the Holocaust in Germany (Herbert

1998). Following the German publication of Goldhagen's book *Hitler's Willing Executioners* in 1996, the public debate in Germany about the country's past changed dramatically. Although the book met with sustained criticism from professional historians, Goldhagen has been praised for – at last – putting the mass murder itself, the motives of those who killed and the suffering of the victims right at the centre of research. Questions which had been asked by historians before – whether the Holocaust was to be interpreted as a phenomenon of modernity, whether it was a putative self-defence of the bourgeoisie against the unconscious wish to kill the Bolsheviks or the issues in the highly abstract debate of the late 1970s about fascism and totalitarianism inevitably paled against the empirical facts of an unimaginable brutality. Goldhagen's book, whatever its scholarly shortcomings, became a milestone by breaking an unwritten societal taboo. It describes in hitherto unheard and unread detail the horror of the actual murders committed. By contrast, traditional Holocaust research had focused on the genesis and consequences of the mass murder, while playing down – for pietistic reasons or to avoid criticism of searching for spectacular effects – what had actually happened in all its concrete and painful detail. Before Goldhagen's book entered the public debate, academic research had been caught up in increasingly theoretical debates about interpretations. They failed to deliver politically transferable answers or any 'explanation' with which one could identify. Goldhagen's account answered the unmet need of the victims and their children by pointing to the clearly defined motives of a large and clearly defined group of perpetrators, while offering to the younger generation of Germans the possibility of siding with the victims. The challenge which was raised and which persists concerns the role played by the 'ordinary' Germans in those murders.

What this case study shows is the subtle, and sometimes not so subtle, interaction of a research field, and its immensely charged political and emotional significance, with the wider society. This cannot be reduced to a simple generational dichotomy (although this remains a persistent theme). Nor can it be compressed into an account of groups in society that differ in their willingness to engage in such a debate or their differences in ideological and political positions, although this also had an impact. What emerges is the unpredictable process of developing a research agenda which results from an intense and painful engagement with different sections of society – with the judiciary, with successive hospital administrations, with the keepers of bank archives or – reaching deeply into personal loyalties – with former academics or even currently respected professors whose role in trying to shape the 'demographic policies' of the regime has come to the fore only recently. The actual development and shape that the historical research agenda took in this case can only be understood by following the path of questions raised and answers provided in highly selective ways (which, in retrospect, always contain 'blind spots' and 'deficits' that are filled only later).

The responsiveness of these different sectors of society – both in a positive (affirming and open) and negative (denying and avoiding) sense – to streams of attitudes, political issues, media reports and judiciary proceedings shaped the public debates at each period. There is one step further to go. Holocaust research cannot be conceived without these continuous layers and series of painful interrogations and intense, emotional interactions with different groups in society, both within one country and abroad. The 'context of implication' in shaping the knowledge and historical findings thus produced is central – but it cannot be planned, or foreseen. It is as much an unintended, long-term process as history itself, which in retrospect can be seen to have a direction. Although it is unplanned it is none the less co-produced by successive generations of historians in the new questions they raise and the discourse they must enter and maintain with society at large.

Can something similar be imagined to occur in a research field in the natural sciences? What is the nature of the influences, for example, which impinge on the development of genomic or post-genomic research? While historical research and much scholarship in the social sciences and humanities might be expected to be open to 'outside influences' that lead to a necessarily close involvement with the context of implication, can this also happen in the natural sciences? Does the absence of people as scientific objects of investigation, at least as a rule, tend to 'protect' the natural sciences from too close an entanglement with how research questions are to be posed or how society may 'influence' what happens on a deep, epistemological level? In the post-genomics era, some scientists are explicitly seeking to address historical, technical, epistemic and cultural aspects of genome projects and are developing arguments to show why closer attention should be paid to the epistemological aspects – and not only to ethical, social and legal issues – of the new genomic mode of practising biology (Collins et al. 1998). These altered practices include an assessment of the new ways of processing and delivering data, new forms of co-operative projects, new relations between knowledge production and application, and of the lessons gleaned from developmental biology, of epigenetics, of the reintroduction of morphology, and of whole-animal representation in conjunction with genomics. The focus here is on the inclusion of the complete dynamics of the organism in post-genomic research rather than on mapping only the genome – a move which would re-configure what it means to do biology in the coming decades.

While we do not intend to compare what it means to do biology with doing historical research in a politically and emotionally highly charged field, the question nevertheless poses itself of how to analyse processes of contextualization that occur in various scientific practices and through shifts in research questions on a deeper level. The manifest presence of a wide range of scientific practices leads one away from a fixed epistemological core operative throughout all of science to a multiplicity of shifting scientific practices. Why practices shift, in which directions and driven by what, remains to be explored in each case. But we should not be surprised to find a multitude of constraining and driving factors which forcefully demonstrate that something is to be gained by conscientiously attempting to understand the web of wider interactions and their consequences on the actual practices, the questions asked and new directions to be explored. If, in the conference on post-genomics mentioned above, the prospect was raised that the epistemological core of biological experimentation is about to change – from being hypothesis-driven, based on the results of single experiments, to being based on systematic, large-scale data production, followed by modelling – this might also be interpreted to show that the contours of an altered context of implication are becoming visible. Such a possibility should then provide an appropriate theme for comparative and historical case studies that evaluate these changes against the broader contours of (not only the life) sciences and their expanded horizon of middle-range contextualization.

11

From Reliable Knowledge to Socially Robust Knowledge

Our argument so far has passed through three stages. First, we have argued that the great conceptual, and organizational, categories of the modern world – state, market, culture, science – have become highly permeable, even transgressive. They are ceasing to be recognizably distinct domains. As a result, common-sense distinctions between the 'internal' and the 'external' are becoming increasingly problematic, a change which has radical implications for demarcations between science and non-science and for notions of professional identity and scientific expertise. We have then focused more closely on the value systems and institutional and professional structures of knowledge production, and argued that, just as the boundaries between state, market, culture and science are becoming increasingly fuzzy, so too are those between universities, research councils, government research establishments, industrial R&D, even other knowledge institutions (for example, in the mass media and the wider 'cultural industries').

Second, we have emphasized that the range of external factors that scientists must now take into account is expanding – inexorably and exponentially. This is hardly a novel or a controversial observation. The increasing emphasis on the contribution of science to wealth creation (and social improvement), the growing deference to so-called 'user' perspectives, the great weight now attached to ethical and environmental considerations, are all examples of the intensification of what we have called contextualization. But, in the eyes of many scientists, this contextualization is unwelcome and imposed. It is something that must be lived with – and worked round – because it is only by acknowledging these apparent 'externalities' that the funding base of science can be secured. In the light of the first stage of our argument, that is clearly not a view we share. In a fuzzy world contextualization cannot be reduced to an accumulation of 'externalities' – which cannot properly be characterized as such. We have also argued that a real shift has taken place from weakly contextualized to strongly contextualized knowledge production; this is no mere rhetorical shift from a once dominant Mode 1-like discourse of, say, physics, to a Mode 2-like discourse of, say, bio-engineering. Again this assertion is not especially controversial. There is substantial evidence, anecdotal and statistical, to back it up.

So, third, we have argued that under contemporary conditions the more strongly contextualized a scientific field or research domain is, the more socially robust is the knowledge it is likely to produce. What does that mean in practice? First, social robustness is a relational, not a relativistic or (still less) absolute idea. For example, the strength of a building depends on a wide range of factors – its construction materials and methods, its physical environment, its social uses and so on. It follows that the social robustness of knowledge can only be judged in specific contexts. Next, social robustness describes a process that, in due course, may reach a certain stability. Third, there is a fine but important distinction to be drawn between the robustness (of the knowledge) and its acceptability (by individuals, groups or societies). Of course, the two are connected – but social robustness, in an important sense, is prospective; it is capable of dealing with unknown and unforeseeable contexts. Fourth, robustness is produced when research has been infiltrated and improved by social knowledge. Fifth, and last, socially robust knowledge has a strongly empirical dimension; it is subject to frequent testing, feedback and improvement because it is open-ended.

However, this is a reversal of the traditional pattern of scientific working, which has been to restrict as far as possible the range of external factors, or contexts, which must be taken into account. Many of the most powerful scientific techniques – reductionism, normalization, sampling methods, control groups – are based on this presumption of containment or insulation. The laboratory, or wider research arena, has been a sterile space – in a metaphorical as well as physical sense. Good science has been constantly at risk of being contaminated, even overwhelmed, by a surfeit of contexts. Our argument is that this has now been turned on its head. Those scientific fields which have continued to restrict the range of external factors which they take into account, to preserve a 'sterile space', and which we have characterized as weakly contextualized, are tending to become less creative and productive. Those which embrace, willingly or otherwise, a diversity of external factors, and which we have described as strongly contextualized, are not only more 'relevant' (this is an inevitable outcome, whether welcomed or resented), but may also be more successful in terms of both the quantity and the quality of the knowledge they produce.

In the next two chapters we move on to a fourth, and crucial, stage in the argument. Some, perhaps many, scientists may be prepared to accept that more highly contextualized knowledge is not only inevitable (in the sense that, in the absence of contextualization, resources for science would dry up) and 'relevant' (in the sense that political, economic and social agendas are more directly addressed) but even scientifically beneficial (in the limited sense that a wider range of perspectives and techniques may be brought to bear on scientific problems). But most will be reluctant to admit our contention that the more highly contextualized the knowledge, the more reliable it is likely to be – not necessarily within the reductionist framework of disciplinary science, which defines reliability almost exclusively in terms of replicability, but because it remains valid outside these 'sterile spaces' created by experimental and theoretical science, a condition we have described as 'socially robust'. We will attempt to demonstrate this larger reliability of contextualized knowledge. In doing so we will question some of the epistemological assumptions on which science is thought ultimately to depend.

Reliability is considered the major epistemic value of science. Without reliability there is no science. The most sophisticated models of the natural world would be useless if they could not be relied upon to be correct (at any rate within the constraints of the available science). As a result, scientists have developed highly elaborate procedures and methods for testing, cross-checking and validating results and theories in order to produce what approximates to 'good science'. Of course, scientists are not alone in their commitment to reliability. Lawyers, accountants and many other professions are equally concerned that the outcome of their work should be 'correct'. However, there is an important difference. Scientists test their results 'against Nature', not against books of rules such as laws or against a body of figures and procedures designed to produce a balance sheet. A more fundamental difference is that the search for reliable knowledge is embedded within the basic belief system of science, both conceptually and in terms of its empirical practices; it is not an externally imposed requirement or constraint. Although mistakes and faults may go unnoticed initially, eventually they will be discovered by other scientists in new contexts. This compulsion continuously to check and test one's own and each other's claims and results is deeply ingrained in the training of researchers. It has also been institutionalized in scientific practice; a good example is the pervasive peer-review system. There is a constant fear of contamination – whether by 'natural artefacts' that invade the experimental environment, producing 'dirt' or causing 'noise', or by the intervention of social, economic or political interests which are also suspected of distorting the reliability – or, more fundamentally, the truth value – of scientific results. It is for this reason that the commitment to the autonomy of science and the belief that it must be independent from other social institutions and systems are so strong. In the eyes of many of the institutional leaders of science, any penetration of science by other cultures – whether democratic or commercial – is bound to compromise its autonomy and, therefore, must be resisted (Ziman 2000).

However, it is clear from the work of Lorraine Daston, Peter Galison and other historians of science that the search for absolute, or near-absolute, 'truth' was gradually replaced by the more pragmatic goal of producing 'reliable' knowledge (Daston and Galison 1992). Science's assertion of intellectual, and institutional, autonomy has been justified – modestly – by the need to maintain the conditions under which such knowledge can be produced rather than – arrogantly – by a claim to reveal or recognize 'truth'. Indeed, science's autonomy is rooted in uncertainty, the need in the absence of terminal 'truths' to maintain disinterested (and, therefore, uncontaminated) methods by which all theories, results and other claims can be constantly interrogated. So what emerges from the history of science is a gradual process of complex objectification rather than grand assertions of the discovery of once-and-for-all methods guaranteeing objectivity. Moreover, the concept of objectivity itself is a many-layered, polymorphous and polyvalent notion. It comprises many strands – moral, methodological, epistemological and pragmatic-instrumental, the cumulative effect of which is to create the conditions of objectivity. But the meanings attached to objectivity do not cohere either in precept or in practice. Sometimes objectivity is seen as a method of understanding, the stepping back from a subjective point of view, at other times as an attitude or ethical stance that can be described as neutrality or impersonality. But whatever its many-layered meanings, it is still the best method to obtain certified and reliable knowledge. Without this reliability, science would only be a game of the imagination, a powerless effort leading nowhere.

No one has done more than John Ziman to emphasize this aspect of science. In his view, science produces reliable knowledge because of the particular rules which all scientists must adopt in the process of generating and communicating new knowledge. For Ziman, '(s)cientific knowledge is the product of a collective human enterprise to which scientists make individual contributions which are purified and extended by mutual criticism and intellectual co-operation' (Ziman 1991: 3). In its simplest form, the model consists of a number of independent researchers, linked by a common set of processes governing the status of the knowledge produced. According to this model, the goal of science is to extend the sway of rational opinion over the widest possible field. Furthermore, Ziman argues that scientific knowledge can be distinguished from other intellectual artefacts of human society by the fact that its contents are consensible; namely that each message should not be so obscure or ambiguous that the recipient is unable to give it whole-hearted assent or to offer well-founded objections. To be comprehensible to others (although the definition, and so the extent, of 'others' is crucial) is necessary if a scientific object, whether theory or set of empirical findings, is to be available to be added to the stock of knowledge or open to improvement or refutation.

Second, according to Ziman, the goal of science is to achieve the maximum degree of agreement – that is, of consensuality. Ideally, the general body of scientific knowledge should consist of facts and principles that are firmly established and accepted without serious doubt, by an overwhelming majority of competent, well-informed scientists. Thus, 'it is convenient to distinguish between a consensible message with the potentiality for eventually contributing to a consensus, and a consensual statement that has been fully tested and universally agreed.... (I)ndeed, consensibility is a necessary condition for any scientific communication, whereas only a small proportion of the whole body of science is undeniably consensual at a given moment' (Ziman 1991: 6). It is through the operation of the twin processes of consensibility and consensuality that science is able to produce reliable knowledge.

If messages are so obscure that they cannot be understood and hence open to being tested and purified by others, then consensus cannot be established. It follows that such messages should not be admitted to the realm of science. Equally, if the social relations of knowledge production and economic conditions under which it takes place inhibit the free flow of information, this crucial consensus-building process is undermined and, with it, the claims of science to produce reliable knowledge. Such an argument is hardly novel; it has been used for many years, for example, to protect science from being co-opted by, or subordinated to, the values of business and political cultures. Business culture, in particular, because of its focus on short-term performance and its ethos of confidentiality stimulated by the perceived need to protect intellectual property, is often criticized for its predisposition to distort the free flow of information. Such distortion restricts consensibility and weakens the fabric of consensuality, the very elements which, Ziman and many others argue, enable reliable knowledge to be produced.

However, there are two main difficulties within this model – which, admittedly, has been presented in a perhaps oversimplified form. The first, conceded by Ziman, is that, although it is science's goal to achieve the maximum degree of consensuality, nevertheless only a small proportion of the whole body of science is undeniably consensual at a given moment. But, if the objective of maximum consensuality can only be achieved sometime in the future (and a future that always recedes), what is its current status, and force? In practice most scientists, in terms of their day-to-day behaviour, seem content to work within the very limited consensus(es) of their specialisms. It is within these relatively limited horizons of discipline-bound peer-group consensus that research activity takes place, theories are constructed and evidence collected and analysed, and scientific reputations are established and maintained. Only rarely can an overarching group of 'integrators' be identified who are attempting in a systematic way to extend these specialized consensus(es) to the 'maximum extent'. Nor does it seem to be true that scientific recognition by peers depends on one's findings being absorbed into the existing consensus. Rather, that consensus may not arise until much later when all the implications of a particular theory or set of findings are corroborated. Even Nobel Prize winners who have launched new research trajectories often have to leave the building of the 'maximum range of consensus' to historical processes. In terms of consensus, then, the ideal of a widespread consensus seems to lie always in the future, often the remote future. In the short term, the consensus-building process is a much more local affair determined by the peer group of each specialism.

The second difficulty, which follows from the first, is that the emergence of ever finer specialisms within science leads to a proliferation of ever narrower consensus(es). Again, this ever finer delineation of scientific domains apparently contradicts the thesis that major societal sub-systems are being de-differentiated; but these phenomena operate at very different levels. But, if, as the Ziman model of consensuality and consensibility suggests, the more widespread the consensus the more reliable the knowledge, how can reliable knowledge be generated when most disciplinary experts frankly admit that they know more and more about less and less? Rather the disintegrative characteristics of modern scientific practice appear to be reducing the prospect of producing reliable knowledge – if its reliability rests on at least the possibility of achieving widespread consensus. The whole process appears to have been thrown into reverse. Of course, someone trained in the techniques of a particular specialism, in principle, could check research findings in another specialism – but hardly so in practice. This cross-disciplinary cross-specialism checking capacity depends on two conditions being satisfied – first, that generic techniques can be identified and applied (apart from allegiance to common norms of behaviour); and second, that the consensibility of scientific findings in different disciplines and specialisms can realistically be maintained.

Neither condition is easy to satisfy. Instead scientists rely on the trust they have in each other and in the integrity of the consensibility process. Not only do they trust their colleagues; they also trust that the publication of research findings which enables them to be checked, and improved or refuted, inevitably brings to light 'wrong' facts and eliminates unsustainable theories, even within the narrow territory of a particular specialism. But, again, this only happens in the long term, which raises again the first difficulty. So instead of there being a broad consensus, the pattern that emerges is of a set of discrete consensus(es). Although each is the product of the dynamics of research within a specialism, severely limiting the possibility of creating real consensus, consensuality nevertheless is maintained by the exercise of mutual trust among scientists, based on their belief/agreement that the process of open interrogation of research findings is broadly similar across specialisms. There is another reason for believing that the emergence of more and more specialisms and sub-specialisms does not threaten the long-term production of reliable knowledge. It can be argued that, so long as scientists are free to pursue their own ideas, they will inevitably raise problems that cross fixed disciplinary or sub-disciplinary boundaries. As a result the unconstricted pursuit of knowledge, through a vast process of checking and cross-checking, in the end, will still produce a robust fabric of reliable knowledge. The very proliferation of discipline and specialism-bound consensus(es) will ensure their intermingling, so producing a web of overlapping consensus(es) which arguably approximates to Ziman's grander overarching consensus and certainly produces the conditions for consensuality.

These processes of consensibility and consensuality admittedly are slow. However, they are self-regulating and systematically eliminate errors. It is argued that these processes are most effective if the science system is kept separate from other forms of knowledge production which might transgress the imperatives of consensibility and consensuality. The capacity to create consensus depends on the clarity with which research results are communicated, which in turn depends on systems of training and socialization, which in their turn can only be expressed through the practices and procedures of the particular specialisms into which, finally, the modern scientific enterprise is currently sub-divided. It is an apparently convincing chain of argument. Consensibility and consensuality are now global phenomena which sustain the belief in the universality of scientific knowledge, although, at any particular time, this universality only comprises an aggregation of the relatively constricted consensus(es) of an ever increasing number of specialisms.

None the less, the actual operation as opposed to the principled articulation of consensuality reduces the potential for producing reliable knowledge. If, according to Ziman's model, the achievement of the greatest possible consensus remains the long-term goal, the reliability of the knowledge actually produced (that is, in the short and medium term) must be compromised by the difficulty of creating anything apart from a plethora of constricted, and perhaps incommensurable, consensus(es). And there is a further complication; it is now routine to regard the corpus of scientific knowledge as provisional, reflecting the prevailing pattern of specialist consensus(es) which are subject to rapid change and radical re-configuration. To the extent that these consensus(es) are provisional their reliability is reduced – but to what extent and in what ways cannot be established. It would be wrong to push the argument too far by inferring that the increasing specialization of science is producing increasingly unreliable knowledge; the provisionality of science from which the volatility (and, arguably, unreliability) of scientific consensus(es) is derived is a reflection of its dynamism which, in turn, reflects science's more sophisticated understanding of the natural and social worlds – which is itself a form of enhanced reliability. However, there must be some relationship between the breadth of the consensus and the degree of reliability that can be claimed, although discipline-bound research is not inherently more reliable. A limited consensus within a particular specialism involving a single peer group may be very similar to a limited consensus in a cross-disciplinary field involving different peer groups in terms of the reliability of the knowledge it comprises. In both cases the consensibility criterion applies; research outputs must be sufficiently comprehensible to be tested. But the locus of consensuality will be different because the composition of the relevant 'experts' is different (Rip, Misa and Schot 1995). None the less a similarly limited consensus is achieved – arguably to the detriment of the reliability of knowledge produced.

In brief, science produces reliable knowledge because, if the rules which guide research practice are followed, a limited and provisional consensus is produced. But not everyone is equal. It is only important that such a consensus is established among the members of a tightly knit community of other specialists, within a particular peer group. They alone, it is argued, are in a position to judge their peers and to uphold the standards leading to good science. In the context of the development of Mode-2 knowledge production, with its shift from a discipline to a problem focus, a key question, therefore, arises. In this new environment characterized by more intense interaction within a much wider community of other practitioners – embracing other disciplines, or stretching across the boundaries of academia and industry or embracing even more heterogeneous 'stake-holders' – is it still possible to produce reliable knowledge? In *The New Production of Knowledge* we acknowledge that the question of how scientific quality could be maintained in a Mode-2 environment had not been properly addressed. To argue that, in principle, additional criteria have to be considered in addition to traditional scientific excellence, even if it continues to be a predominant element in the quality control system, is easy enough; the tensions, dilemmas and even contradictions that actually arise in scientific practice are more difficult to dismiss. So it is essential to consider how a shift to Mode-2 knowledge production, to greater emphasis on problem-relevance and on the specific contexts of application in which this relevance arises, affects the formation of consensibility and consensuality.

The position of Ziman, among others, is that the closer the interaction between academia and industry, the more difficult it becomes to sustain the traditional academic ethos – and, in particular, the commitment to disinterestedness which marks the frontier between academic science and industry. Mode-2 knowledge production, with its emphasis on problem-solving in the context of application, tends to erode the independence of academic researchers and, consequently, weakens their defences against external influences. These influences are not hypothetical. The frontier land between academia and industry is rapidly becoming populated by semi-autonomous research entities that earn their living by undertaking specific projects. These projects are supported by a variety of funding bodies, including private-sector firms and government departments. Some of these bodies make elaborate efforts to foster originality and integrity; after all, it is hardly in their interest to commission second-rate research. But inevitably there is a general tendency, and often explicit instructions, to favour projects with apparently better prospects of 'wealth creation', or with practical, medical, environmental or social applications. Mode-2 research, however remote from actual application, is characterized by its potential for use. Ziman's verdict is unequivocal: 'Although Mode 2 may also incorporate traditional scientific values – including, of course, the sheer obduracy of physical reality – it is clearly an activity where socio-economic power is the final authority' (Ziman 1998: 145).

But does this applications orientation and engagement with 'extra-scientific' forces fatally undermine Mode 2's capacity to produce reliable knowledge? Certainly secrecy and short-termism are counter-productive. If sustained over a long period, they would seriously weaken reliability. But how precisely does this broader configuration of researchers coming from universities and industry, government laboratories or even the semi-autonomous research entities referred to above, who have been orchestrated or 'self-organized' to search for solutions to a common problem, undermine consensibility and consensuality? Consensibility, as has already been explained, requires the transmission of clear messages about the research questions to be addressed, techniques employed and results obtained. It is not clear why research carried out in a problem context should weaken consensibility within the problem-solving group. In fact, if clear communications are not maintained within the research group, less satisfactory solutions to problems are likely to be produced. Consensibility, although it may be more difficult to sustain under Mode-2 conditions, is equally – or perhaps even more – crucial. It can even be argued that, because communication within a Mode-2 research group cannot be taken for granted (which, sometimes mistakenly, might be the case in a tightly knit Mode-1 group), greater emphasis will be placed on the need to create consensibility.

In a similar way, consensuality would still operate under Mode-2 conditions even if a more diverse set of people with different scientific backgrounds and training, and probably with different attitudes and outlooks regarding the potential of future use, has to be engaged. It is not clear why the importance of consensuality should be less within such a diverse group, despite the (apparently) greater difficulties in communicating and reaching a consensus. As discussed in chapter 10, Peter Galison has described the emergence of so-called 'trading zones' characterized by ingenious processes of negotiation between the different parties. For example, they devise 'pidgin' languages which bridge disciplinary gaps, and allow everyone to understand and be understood by the others. The members of such groups have very clear and nuanced strategies of mutual exchange at work because, in their pursuit of a common objective, they depend on the knowledge, skills and know-how of others which they would be unable to obtain otherwise (Galison 1997). But there is no suggestion that consensuality, like consensibility, cannot be practised in such a context.

However, even if this is accepted, two further issues must be addressed – the potential for replicability and the impact of secrecy. The first presents fewer difficulties. The results of Mode-2 research, at any rate in principle, can be checked and verified by others outside the research group, or the problem context, provided the conditions of that context can be replicated. Precisely the same replicability requirement applies to traditional experimental science. Nor is the difficulty of replicating these conditions necessarily greater in a Mode-2 environment. Replicating scientific findings even within a disciplinary context has never been straightforward, because the same (or similar) experimental conditions must be replicated, which is rarely easy. In empirical terms it is difficult to draw a categorical distinction between Mode-1 science and Mode-2 knowledge production in terms of ease of replicability and, therefore, of potential for verification.

Nor does secrecy present insurmountable difficulties to preserving consensuality and consensibility. Indeed, it can be argued that the threat to consensibility apparently posed by secrecy (because it inhibits the free flow of research messages) may even be greater in the absence of a common problem-solving context. Under Mode-1 conditions, knowledge may be more detached and prone to distortions because it is produced in an environment remote from its eventual applications. But under Mode-2 conditions when supply and demand are articulating one another, when researchers move from one problem context to another and when the composition of such research teams varies over time, secrecy is clearly dysfunctional because, except in the very short term, it places more obstacles in the way of producing effective solutions and, therefore, is counter-productive. The challenge is to gain access to knowledge and expertise, which are often widely distributed, rather than to protect intellectual property (because scientific protectionism makes it more difficult to gain access to the intellectual property of others).

But, even if there are no principled reasons why consensuality and consensibility cannot be maintained under Mode-2 conditions, clearly it would be misleading to suggest that nothing has changed. Something must be different, even if the outcome can still be called reliable knowledge. The question must be asked: does Mode 2 produce a different kind of reliability and if so, in what sense? Reliable knowledge, as traditionally defined, presumes that consensibility and consensuality operate mainly within a disciplinary context, even if disciplines are constantly forming and re-forming. The relevant community is easily recognizable, its knowledge is specialized and peers are known to each other. The reliability of such knowledge is rooted in its relevance to these disciplinary or sub-disciplinary structures; it is inner-directed. In contrast, reliable knowledge produced by a wider network of collaborators working under Mode-2 conditions is reliable in terms of a wider consensuality; it is outer-directed. It is reliable in terms of the problem relevance of the context in which it arose and which continues to influence it. Of course, these two kinds of reliability are not separated from each other in water-tight epistemological or practical compartments. There are visible and invisible, formal and informal cross-currents and systems of exchange between them. To occupy one of the many 'trading zones' in which inter- (or trans- or even intra-) disciplinary knowledge is created does not require the abandonment of one's disciplinary 'home', or the loss of one's primary identification as an academic or industrial scientist. In some respects it may even strengthen disciplinary and professional identities. In other respects subtle accommodation processes may be at work, creating multiple and modified identities which are different. The experience of life in a 'trading zone' may be not unlike that of the emigrant (although some trading-zoners will return to their old disciplinary 'countries' – and maybe return again, perhaps frequently); old loyalties are fiercely maintained while new identities are created, producing new hybrid-researchers.

Reliable knowledge, therefore, has always been reliable within bounds. In its conventional (and constricted) and refined (or 'pure') sense, these bounds embrace a relatively small number of peers. They police the bounds of reliable knowledge, ideally through disciplinary cohesion, in order to limit as far as possible the contamination likely to be produced if the surrounding social context is allowed to enter the realm of knowledge. The ethos of academic science, and its social practices, were designed to achieve this containment. But it is difficult to see how this containment strategy can remain valid, except in the exceptional context of the need to preserve scientific 'enclaves', when science has entered the *agora*. In this new environment the boundaries within which reliable knowledge is contained have been dramatically extended, even abolished. Reliable knowledge, as validated in its disciplinary context, is no longer self-sufficient or self-referential. Instead it is endlessly challenged, and often fiercely contested, by a much larger potential community, which insists that its claims to be heard are as valid as those of more circumscribed scientific communities and demands that its preferences, too, be taken into account. Objective knowledge is no longer sufficient in itself if the outcomes of its objectivity are ceaselessly negotiated, contested and even rejected. Of course, reliable knowledge in terms of its disciplinary relevance and validity, like objective knowledge, is not simply to be discarded or ignored. Reliable and/or objective knowledge continues to provide the foundations on which our knowledge of the natural world depends. But neither is any longer sufficient of itself.

Otto Neurath repeatedly used the powerful metaphor of scientists having to rebuild their ship while on high seas (Cartwright et al. 1996). This metaphor has generally been interpreted as referring only to the cognitive content of science. The new planks needed to replace the old and decaying wood have to be taken from other parts of the ship, while ingenious *bricolage* and engineering help to keep it not only afloat but also ploughing through the waves. However, in the light of Neurath's lifelong passionate interest in how science could reach out to a wider public, he would probably have accepted a much wider interpretation of his metaphor. This metaphor – of scientists rebuilding the ship of knowledge while it forges ahead – can also be used to describe the more heterogeneous (and also more ambiguous) constitution of reliable knowledge now needed to keep the ship of science afloat amid the turbulent waters of contemporary society. To suggest that reliable knowledge must engage the social world more openly and directly is not to seek to diminish but to enhance its status and validity, by arguing that reliable knowledge – to remain reliable – has also to be socially robust knowledge.

12

The Epistemological Core

The most radical part of our argument, and therefore the most difficult to accept, is that the co-evolutionary changes discussed in chapter 3, which we have bundled together under the convenient label 'Mode-2 society', the transformation of knowledge institutions, pre-eminently the university itself, described in chapters 5 and 6 and the contextualization of science analysed in chapters 7–10 have made it necessary not only to re-conceptualize the reliability of knowledge but also to question its epistemological foundations. Our contention, contentiously, is that the epistemological core is empty – or, alternatively and perhaps more accurately, crowded and heterogeneous. That irreducible core of cognitive values and social practices, which once enabled good science to be distinguished from bad science (if not – quite – truth from untruth), has been both invaded – by forces once defined as extra-scientific – and dispersed, or distributed, across more, and more heterogeneous, knowledge environments. But it is the 'success' of science that has produced this result, the invasion and distribution of the epistemological core. Our argument, therefore, is that a more nuanced, and sociologically sensitive, account of epistemology is needed.

The rise of uncertainty, which we identified as one of the most powerful co-evolutionary trends in Mode-2 society, was compounded at the end of the twentieth century by an inevitable revival of millenarianism. Even in a thoroughly secular society the temptation to identify the multiple transformations of science and culture, society and the economy, with the shift to a new millennium appears to be irresistible, however atavistic such identification may appear. So it is not surprising that our view of science in the twenty-first century is associated with grand phrases such as 'cognitive revolution', 'information explosion', 'global economy', 'labour flexibility', 'environmental sustainability', 'wealth creation' and so on. These phrases reflect science's boundless authority, limitless expectations and inexorable applications on the one hand and on the other its identification with grand, and even mysterious, transformations in a millenarian mould. The association of science with such diverse phenomena also indicates that science is no longer seen as occupying an autonomous – or, at any rate, discrete – domain. Instead it has become a transgressive category spreading into other domains (as a result of its 'successes'?) and, in turn, being absorbed into them (as a result of its 'failures'?).

These phenomena suggest some of the objectives, boundary conditions and opportunities shaping the future of science. To take just two examples: developed nations are more determined than ever to exploit science's (and technology's) potential for innovation in order to secure competitive advantage in the global economy; while global corporations are determined to exploit the developing world's potential, whether as a source of cheap labour or as an emerging market. Science and technology, irreversibly identified with the modernization project, are deeply implicated in these wider socio-economic processes which cannot be envisaged without their intervention and assistance. Scientists have become indispensable facilitators in the realization of all kinds of societal objectives, many of which cannot be categorized even approximately as 'scientific'. They have also become the protagonists – and antagonists – of powerful social, economic and political interests. Although scientists still assert their claims to institutional, professional and cognitive autonomy, they no longer occupy an autonomous space. To the extent that science has become embroiled in the realization of extra- (or supra-) scientific objectives its 'space' has been invaded, or compromised, and its boundaries permeated by these other objectives which are linked to other social, economic, cultural and political 'spaces'.

The emergence of Mode-2 society will inevitably lead to further invasion of the 'space' occupied by science and greater permeability between this and other 'spaces'. The societal objectives designed to realize the scientific-technological potential embedded in an expansive twenty-first century conception of science are many and contradictory. Their diversity, and contradictions, raise fundamental questions. First, how easily and quickly can social institutions cope with scientific-technological innovation, and under which social conditions does such innovation flourish or is it hindered? Second, technical arguments about path-dependence in technological regimes and patterns of economic development have been developed into broader assertions that 'history matters' in the wider context of economic change and that, as a result, distinctive institutional legacies generate contrasting outcomes in different societies. Third, it is argued that differences between the generation and diffusion of innovation in different business systems can be attributed to distinctive patterns of funding and methods for controlling public research. If, for example, the pattern of innovation in German and British companies, reflecting distinctive growth and risk management strategies, is different, to what extent can these variations be explained by more general characteristics of their business systems and to what extent by the organization of research in these countries? (Whitley 1999). Even in the case of 'wealth creation', to which all are assumed to subscribe, contradictions abound. There are complex debates, in both a technical and an ethical sense, about the need for re-distribution, about social justice and economic equity, about strategies to protect vulnerable members, and classes, of society from the harsh forces of economic globalization. Instead of simple answers there are only partial solutions, temporary accommodations between different material interests, social values and ethical visions. Despite science's potential for innovation the path from present predicaments to viable policy solutions remains long, tortuous and full of unanticipated consequences.

The twenty-first century view of science is the culmination of the continuous 'success' of science and technology. Not only has science literally transformed the world, in both material and conceptual terms; it has also made a key contribution to generating social cohesion and maintaining socio-political consensus, at any rate in democratic capitalist societies, which has often been overlooked; science and technology have fuelled, or seemed to fuel, sustained economic growth and continuous improvement in the standard of living and quality of life. If Western society has enjoyed exceptional stability in the past half-century this is largely due to its material success, a record which few other forms of society have been able to match; this success, in turn, appears to have been intimately linked to scientific and technological innovation. This triumphant calculus has produced socio-economic progress and scientific and technological success, often seen as two sides of the same coin, which has far exceeded the most ambitious claims and predictions of the Enlightenment. However, amid this success, two decisive breaks have taken place. It is not simply that the 'risks', or 'burdens', of progress have become increasingly visible alongside the successes and benefits that such progress has undoubtedly produced and, therefore, that these unanticipated, and undesirable, consequences must be managed and minimized by means of foresight exercises and technology assessments. Nor is it simply that critiques, doubts, refutations – which we have bundled up under the heading of 'contestation' – have proliferated in many fields of science and technology and, as a result, have become part of the routine interaction between science and society. These breaks with the onward-and-upward 'success' of science, and their causes, go much deeper.

The Ambivalence of Novelty

The first break has been produced by the revival of a deep-rooted, almost atavistic, ambivalence about the unparalleled success of science and technology in stimulating novelty. This ambivalence reflects the inherent uncertainty that novelty, and indeed every innovation, brings with it. Novelty is challenging – and disturbing. It creates uncertainty by not only challenging but also changing established ways of doing things. And not only of doing – also of seeing things, experiencing the world cognitively, conceiving of meanings. The production of novelty and process of innovation are deeply radicalizing. In industry and business uncertainties proliferate; they extend beyond the actual process of innovation to its potential uses and users and to the capacity of innovators to appropriate or to benefit in other ways from their innovations (Whitely 1999). In society uncertainty is produced whenever changes modify social practices, in the arena of action, and subvert social hierarchies, in the context of structure. But scientific novelty and technological innovation are even more radicalizing in their effects. First, they are bundled up with (and often held responsible for) a host of other innovations – social, economic, cultural, political. Users and usages change; traditional authorities are challenged and new ones created; institutions, organizations and other structures are reformed and re-configured. Second, additional uncertainty is produced by anxiety about the impact of novelty and innovation on current relationships, especially when the existing distribution of power, whether economic or symbolic, through these relationships is likely to be disturbed; about who will be winners and losers, and how those adversely affected by innovation will be able to cope with demands to develop new skills and adapt their life-styles. Third, and most important of all, the uncertainties generated by novelty and innovation may threaten our sense of self- and person-hood. Who am I in a transformed world? Am I changed too? When a researcher in a research institute in Scotland linked to a local biotechnology firm successfully cloned the sheep 'Dolly', the subsequent public unease was produced not by concerns about the planned application of this innovation as a drug incubator. Rather the anxiety arose because 'Dolly' appeared to highlight the vulnerability of our sense of human individuality and identity.

Enthusiasm and uncertainty go hand in hand. We embrace the new but fear those of its uncertain consequences which, because they cannot be anticipated, are difficult to control. This ambivalence towards novelty, of course, has always existed. But it has been intensified by the 'success' of modern science because both enthusiasm for its boundless potential and uncertainty about its unknowable consequences have been intensified. Understanding and doing, representing and intervening, knowledge and power, have moved together in intractable ways. They have almost become synchronized, fuelling fears that everything which becomes known will inexorably be implemented and that everything discovered in the laboratory will be quickly applied in the outside world, provided that a market (whether commercial or political) can be found. The gap between knowledge and action has been narrowed to such an extent that they cannot be separated, temporally or organizationally. This combination of knowledge and action is a far more radical process than the ever increasing velocity of innovations. It has produced two results. First, the uncertainty created by scientific novelty is no longer a future event, a distant foreboding, something which is amenable to subsequent correction. Instead it is a present experience, an instant phenomenon that is integral to the initiating novelty. Second, the instantaneity of novelty-uncertainty provokes resistance. It is as if, conscious of the infinite variety of our fantasies and desires and of science's ever increasing capacity to satisfy some of them while stimulating ever new ones, we recognize the need to reassert control. As Paul Rabinow writes about the Human Genome Project: 'the object to be known – the human genome – will be known in such a way that it can be changed. This dimension is thoroughly modern; one could even say that it instantiates the definition of modern rationality' (Rabinow 1992: 236).

This elision of discovery and action – or, more accurately, a proliferation of actions – means that the work of scientists has acquired intense social significance. It has moved centre-stage in what we call the *agora* – the space in which market and politics meet and mingle, where the articulation of private emotions and meanings encounters the formation of public opinion and political consensus.

The unparalleled success of science in generating novelty has been matched by its infinite ability to create uncertainty. This uncertainty then has to be managed – or, if not managed, then acknowledged – not only in the context of the production of scientific knowledge and its dissemination and application in the market and society, which is difficult but not impossible, but also in terms of the wider societal impact of this knowledge, which is much more problematical. As has already been pointed out, the time between knowing and doing has been radically shortened. As a result, the sequential chain of discovery and application has been broken; we now live in an instantaneous extended present; and the difficulty of making choices is dramatically increased in this foreshortened environment. This can be expressed in neo-Darwinian language: the largely blind variation of what is possible has been greatly enlarged by the success of science, but it has become more difficult to devise mechanisms for selecting what should be retained, accepted, rejected, modified or incorporated. These selection mechanisms cannot be left to science alone; it is in this sense that new and vigorous engagement of science in the *agora* will be necessary. And while it can be argued that the new has always been greeted with ambivalence and a mixture of fear and openness, the time-span to assess its consequences, form a reaction towards it and attempt to accommodate it selectively has shrunk dramatically. Instead of reacting to novelty, a capacity to anticipate what is not yet known is needed. Uncertainty is unavoidable and intensifying; indeed it is the hallmark of the current phase of modernity. As a result innovation strategies are required which are not only multi-layered but also engage the here-and-now as effectively as the there-and-later.

The Decline of Cognitive Authority

The second decisive break in the triumphant success of science is the de-coupling of science's useful outcomes, for which it is now most highly valued, from its cognitive authority. This is a more complex phenomenon than the division of labour, and dependence, implied by the traditional distinctions between pure and applied science, science and technology or research and development. It has more in common with the uncoupling of modernization from modernity, the concrete processes of innovation and improvement from the values on which these processes were once assumed to rely. Here a fundamental paradox is encountered – on the one hand, apparently, an alarming decline in science's ability (and authority) to define the reality of the natural world; on the other hand an unprecedented increase in its power to manipulate and intervene in that world. This paradox is all the more remarkable because scientists, by means of the scientific-technical instruments at their disposal, still have, if no longer a monopoly, at any rate privileged access to ways of understanding and manipulating the natural world. Under these conditions the cognitive authority of science might have been expected to be reinforced. However, the diffusion of a wider range of sophisticated instrumentation into workplace and home, the development of global communication and information infrastructures and networks and the growing emphasis on the potential of user–producer relationships as a key arena for the generation of new ideas and applications on which scientific-technological novelty crucially depends are constantly extending – democratizing and commercializing – access to ways to understand, and therefore manipulate, the natural world. Although we are not all 'scientists' yet, many more of us are.

A first sign that, in the eyes of the public, science's grip was weakening on its near-total monopoly on definitions and explanations of the natural world occurred during one of twentieth-century science's greatest climaxes, the quantum mechanical revolution in physics. Although the educated public lamented the erosion of authority represented by the new physics, it became clear that science would – and could – no longer deliver the same kind of coherent scientific world view that had been produced by its nineteenth-century victory over religion. In that sense the normative-cognitive authority of science was degraded. However, during the next half-century the knowledge produced by quantum mechanics, which appeared to many people abstruse and esoteric, produced a growing number of sophisticated 'quantum objects' which in turn provided the basis for a thriving high-technology industry. In another sense, therefore, the instrumental efficacy of science was enhanced. Paradoxically science's grip on the reality of the natural world began to disintegrate at the very time when its power to manipulate, intervene and alter that reality was increasing.

Various interpretations of this slippage of science's normative-cognitive authority have been suggested which are directly relevant to our assertion that the epistemological core is empty. One interpretation is derived from studies of the cross-cultural transfer of knowledge – especially Japan's exposure to, and exploitation of, Western science in the late nineteenth century. According to Steve Fuller, the Japanese showed that with a little ingenuity domestic forms of knowledge could nearly always be adapted to novel purposes so that the need to import foreign forms of knowledge, and the cultural dislocation they brought with them, could be avoided (Fuller

1997). They developed alternative means to pursue desired scientific ends (which were often identical to those pursued in the West). To achieve this the Japanese followed a variety of strategies; in some cases they exploited their distinctive metaphysical and religious traditions; in others they created hybridized Western institutions and practices. The general lesson to be drawn from the Japanese experience is that the knowledge which survives translation through a series of cross-cultural encounters is simply the residual product of these translations; its persistence does not require any further metaphysical explanation. To the extent that its universalism is compromised or dissipated, the epistemological core can be shown to be empty.

Fuller does not deny the potential value of attempts to standardize, or even to universalize, knowledge across different cultures, but he argues that such attempts are essentially political. For example, standardization may help to establish, or alternatively to dissolve, social hierarchies both within and between societies. In this case standardization is used to promote social cohesion. In other cases attitudes to standardization are shaped by social divisions. For example, producers and consumers of knowledge have different interests. Producers favour standardization; they promote convergent forms of knowledge despite divergent contexts of use because, as a result, consumers have to adjust their interests in order to acquire that knowledge. Their willingness to pay this price is reflected in the attractiveness of educational credentials and the willingness to take, and accept, expert advice. Consumers, on the other hand, favour forms of knowledge that capitalize as much as possible on what they know already and what they want to achieve. Consequently, producers conceive of knowledge as essential, without which their goals cannot be attained; in contrast, consumers see knowledge in economical terms – in the sense that the cost of acquiring knowledge should be reduced to an absolute minimum (Fuller 1997: 132–4). The conflation of producers and consumers characteristic of Mode-2 knowledge production complicates rather than reconciles these competing interests.

According to this interpretation, knowledge has been stripped of its metaphysical and culturally specific elements in the process of cross-cultural transfer. As a result science has been reduced to its utilitarian essence – the cost of acquiring or selling knowledge, whether this transaction takes place across cultural boundaries or across the boundaries that separate scientists from non-scientists or, more generally, experts from lay people. Seen in this light, scientists' assertion of their cognitive authority is nothing more than a 'market' device to increase the price that knowledge consumers must pay, in both the political and commercial arenas, and also to restrict conditions under which these knowledge transactions take place. Fuller describes the standpoint of knowledge producers as 'transcendental'; in other words they conceive of knowledge as a *sine qua non*, without which desired objectives cannot be achieved. This implies that only certain kinds of encounters with, and interpretations of, the social and natural worlds are effective in helping to achieve these objectives, essentially those encounters which can be described as 'scientific'. In contrast, for consumers valid knowledge is not restricted either by *a priori* epistemology or by processes governed by 'scientific' rules. They regard knowledge merely as a good, which is standardized or customized like other goods, and believe that the costs of acquiring these knowledge-goods should be as low as possible.

But does this minimalist account of the relationship between producers and users of knowledge, from which its metaphysical and cultural (and, therefore, authoritative) elements have been stripped out, enable us to conclude that the epistemological core is really empty? It is possible to argue that such cross-cultural knowledge – and, some might even argue, contextualized science – is impoverished by the absence of these elements and that it, indirectly but ultimately, depends on the epistemological core. In other words knowledge is 'transcendental'. If the 'cultures' that are being 'crossed' in cross-cultural knowledge are national cultures geographically separate and historically distinct, this counter-argument has a certain plausibility. Perhaps something important is lost in such cross-cultural transfers, deep cultural associations and affinities which do not 'travel' well even in a global knowledge economy. It may also be that the epistemological core is not so much empty as crowded with 'other', maybe incommensurable, epistemologies. In the case of Japan, Western science, in order to function, had to be associated with new epistemologies; it was not epistemology-free.

But not all cross-cultural transfers of knowledge take place between geographically separate and historically distinct cultures. Many take place between different academic disciplines – and many disciplines have always been constituted as cross-cultural arenas. With these cross-cultural transfers is it possible to argue, as we do, that the epistemological core is empty or, as potential critics may assert, that cross-cultural knowledge is a residue, epistemologically impoverished by its disconnection from metaphysical and cultural affinities? The editors of the French historical journal *Annales* described its influence as 'not that of a school (with the attendant risks of becoming a clique or an institution) or a mailing address (however prestigious) but, rather, a site of experimentation'. Later in the same editorial to mark the journal's sixtieth anniversary they asked: 'Might historians, having taught others to uncover and use historical sources, have nothing left to do but open their territory to an ecumenical practice of the human sciences? The terrain of history might then be abandoned to anthropologists, economists and sociologists of the past.' Their answer, unsurprisingly, was no, because if the economic historian was simply an economist of the past, history and economics would no longer have fundamental truths to teach each other. 'The diffusion of innovations assumes potential differences' (*Annales* 1989: 1317–18).

Images of Knowledge – and the Body of Knowledge

The key ideas here are very relevant to our discussion of the epistemological core. The characterization of disciplines as sites of experimentation suggests not only dynamism but also dissonance, which is difficult to reconcile with epistemological stability. On the other hand, emphasizing 'fundamental truths' embodied in disciplines does suggest the persistence of authoritative structures within them. The argument that innovations are diffused by the potentiality of 'difference' points in both directions – towards the epistemological variation implied by 'difference', and towards the authoritative structures through which 'difference' is sustained. The links between the diffusion of innovation and the persistence of 'difference' also help to explain the apparent contradiction between the decline of science's normative-cultural authority and its growing relevance and utility in the expertise society and knowledge-based economy.

However, there is a second interpretation of why this authority has been undermined, while its instrumental utility has been powerfully reinforced. Nowhere is the loss of the social capital of science, to use Pierre Bourdieu's terminology, more apparent than in its declining authority in the face of public controversies about science. In one sense this is unsurprising; a society that insists on mediating everything else is unlikely to make an exception for science. But the waning trust in science and widening distribution of expertise, both of which have weakened the monopoly of traditional scientific-technical experts, are only part of the explanation. Non-scientists have always interpreted events and experiences, including scientific events and experiences. What may be new is that at no time in history have lay interpretations and part-fictional representations of science been more important. According to Jon Turney, 'the science and technology we ultimately see are partly shaped by the images of the work which exist outside the confines of the laboratory report or the scientific paper. If we want to understand the origin of the vocabulary in which present-day debates about science are conducted, we need to attend, not just to the internal development of science, but to the history of science in popular culture'. Turney goes further, recalling approvingly Mary Shelley's remark about experiments carried out by her great contemporary Erasmus Darwin, Charles Darwin's grandfather: 'I speak not of what the Doctor really did, nor said he did, but, as more to my purpose, of what was then spoken of as having been done by him' (Turney 1998: 3).

This, of course, is unacceptable to many scientists. In the humanities, and even the social sciences, the idea of plural 'realities', or discourses, is commonplace (although still contested); the author is not necessarily authoritative. But among scientists such an idea is profoundly disturbing. Popular images of science, when they are negative, are seen as a threat, and even when they are positive such images are still regarded as a naive, simplistic and, therefore, intolerable distortion of scientific 'truths'. The uncritical popularization of science and its opposite, the growth of an anti-science movement, are both dismissed as irrational. Yet popular images of science represent a vital link between scientific culture and the general culture of society. They help bridge the gap between scientific values and practices on the one hand and, on the other, wider social norms and everyday experience. Seen in this light popular images of science are not only inevitable but also essential for the maintenance of a healthy scientific culture.

Yet these images, by their very nature, blur what scientists call 'facts' with 'fiction', the products of human imagination. It is by means of such blurring that they bridge the gap between science and society. Scientists may struggle to establish, or re-establish, their authority over the demarcation between 'fact', or science, and 'fiction', or non-science (although the possibility can never be excluded that 'fiction' can be transformed into 'fact' through the beneficent operation of a progressive science). Nevertheless it is becoming more difficult for science successfully to assert its right to define objective reality. Popular images of science can no longer be relegated to the status of mere epiphenomena of popular culture by a dominant Enlightenment world view. They can no longer be subordinated and patronized. The question is no longer when folk-science or scientific 'fiction' will catch up with real science or scientific 'fact'. Rather, the question is how the popular images of science shape and interact with science itself. Those who read newspapers, listen to radios and watch television, ordinary men and women in a variety of work and life situations with different educational credentials, are not passive and ignorant recipients of media messages. Instead they work actively to construct and interpret, to make sense of what science is and what scientists do.

A key to this shift is the decay of grand narrative and the growing influence of particularistic local discourses, a familiar enough phenomenon in the wider terrain of social theory but unacknowledged in the narrower context of science studies. 'Fact' and 'fiction' are becoming increasingly blurred because of this shift from meta-narrative to micro-discourses. To lend coherence to grand projects, scientific or social, it has always been necessary to weave them into 'stories'. Scientists, as the shock-troops of modernity, have constructed a grand narrative – in line with the other grand narratives of the Golden Age of modernity, narratives of secularization and rationality, of economic progress and so on. They have attempted to lend coherence and meaning to their project, the enterprise of science, by telling unifying and universalizing 'stories' about the origins of the physical universe, human life and the social world. Science was often depicted in these narratives as context-free, or constituted round theoretical concepts which could safely be assumed to be historically invariant. In these grand narratives there was little room for individual scientists, although the existence of lone and singular scientific geniuses was acknowledged (and their existence proved the rule, which was of depersonalized 'normal science', to borrow Thomas Kuhn's phrase). Science was communitarian. Scientific knowledge accumulated in accordance with processes transcending the mundane work of individual scientists, whose importance lay in their membership of the scientific community. This communitarianism was based on a shared commitment to a single overarching goal, the obligation to develop ideas and test their validity, thus approaching ever closer to an approximation to truth, which was equated with reality.

In contrast, the narratives which validate popular images of science have typically been local and particularistic, embedded in the cultural contexts within which they have been constructed. They are avowedly subjective – in a double sense. First, the sense-maker and his or her problems and world views have been placed at the centre of these images. Secondly, scientists have been seen as individuals, because they deviate from the norm characterizing other individuals who are not scientists. The relationship of any lay person to science is inevitably subjective. The French geneticist and popular writer, Jean Rostand, reflecting in the 1950s on the emotional value of biology, was struck by the correspondence he continued to receive from ordinary people. He concluded: 'the science that provokes such appeals, prayers and confessions, the science that penetrates into private life, and whose warnings or advice can influence a marriage, a decision to have children, a person's destiny, is no ordinary science' (quoted by Turney 1998: 37). Clearly biology is 'no ordinary science', because it touches our most potent and intimate wishes as human beings. But something similar can be said of nearly all other sciences; there are no 'ordinary sciences'.

Ordinary people became agitated by science because their feelings have been aroused by its potentiality. Perhaps they are thrilled, full of admiration for its benefits and in awe of its claims; perhaps they are revolted by its practices or outcomes. But, either way, they are personally engaged. The question they ask is 'what does it mean for me/us?' It is this question, endlessly and incommensurably repeated, that structures the many small narratives devised by the scientifically curious. Their concern ranges from the existential – matters of life, death and ultimate meaning – to the mundane – questions of diet and energy efficiency. Because there are no 'ordinary sciences', popular engagement with science will always remain emotional and subjective. This multitude of many micro-narratives embedded in local and particularistic contexts is utterly unlike, and has few points of contact with, the meta-discourse which science derived from the Enlightenment, the image of a universal and unified 'objective' scientific knowledge.

The guardians of science (in a quasi-Platonic sense) therefore must accept that one part of their job description, custody of science's grand meta-narrative, is becoming obsolete while the other part, articulation of the countless local narratives that compromise the complex links between science and society, is increasing in significance (although these categories are themselves becoming transgressive in terms discussed in chapter 1). There is, of course, an intriguing affinity between the decline of science's macro-narrative and rise of socio-scientific micro-narratives on the one hand and on the other the erosion of science's normative-cognitive authority and its endlessly increasing instrumental utility. The two patterns mirror and merge into each other. Science-as-unified-subject decayed while contextualized science advanced. When science's Golden Age in which an almost Trinitarian relationship between science, the state and industry had become enshrined began to draw to a close, an extraordinary period of the history of modernity was also ending. The hallmark of that period was reliance on orderly systems, and centralization was a crucial organizing principle. The public authority of science was inextricably linked to this predominant values-and-vision system; a belief in the inexorable 'scientization of society' was one of the most powerful glues holding the system together in a rational and rationalized means–ends relationship. The decline of that authority coincided with the blurring of the vision of an orderly system; the state had to abandon some of its previously assigned functions to markets and, as a result, de-regulation and other centrifugal forces became much more powerful. Whether, and to what extent, these forces privileged ordinary individuals as citizens, consumers and producers, to what extent, if any, they have become 'empowered', remains a matter for debate. Although hierarchies certainly became flatter and systems fuzzier, they did not disappear. On the contrary, flatter hierarchies and fuzzier systems may actually be more insidious and therefore more intractable.

In this, and earlier, chapters we have emphasized the growing gap between the actual practices of science and its public imagery. In arguing that the epistemological core of science is empty we have assumed a certain image of science which is still widespread, apparently predominant – and utterly misleading. But it can be argued against us that this is merely an image, not the real thing. Steven Weinberg's recent critique of Thomas Kuhn (Weinberg 1998) concentrated on attacking Kuhn's incommensurability thesis, which had already been comprehensively criticized since *The Structure of Scientific Revolutions* was first published in 1962. But he also argued that it is important to be clear about what changes and what does not change in scientific revolutions, a distinction not made clearly by Kuhn. This second part of his critique is directly relevant to our argument. According to Weinberg, there is a 'hard' part of modern physical theories ('hard' meaning not difficult, but durable, like bones in palaeontology or potsherds in archaeology) that usually consists of the equations themselves, together with some understandings about what symbols mean operationally and about the sorts of phenomena to which they apply. Then there is a 'soft' part; it is the vision of reality that we use to explain to ourselves why the equations work. The soft part does change; we no longer believe in Maxwell's ether, and we know that there is more to nature than Newton's particles and forces.

Weinberg admits that the changes in the soft part of scientific theories also 'produce changes in our understanding of the conditions under which the hard part is a good approximation'. But he adds: 'after our theories reach their mature forms, their hard parts represent permanent accomplishments. If you have bought one of those T-shirts with Maxwell's equations on the front, you may have to worry about its going out of style, but not about its becoming false. We will go on teaching Maxwellian electrodynamics as long as there are scientists. I can't see any sense in which the increase in scope and accuracy of the hard parts of our theories is NOT a cumulative approach to truth' (Weinberg 1998: 50). It is the last part of Weinberg's argument which apparently contradicts the argument that has been developed in this chapter.

This is a clear and concise statement about science's 'hard core' – scientists believe there are truths out there to be discovered which, once discovered, will form a permanent part of human knowledge. This is, Weinberg says, the justification of science's claim to 'have access to reality, in which the laws of nature are real in the same sense as the rocks on the ground – whatever that is' (Weinberg 1998: 52). It is also the justification of science's claim on public resources. To spend on what? – 'to see the theory of gravitation and all of the different branches of elementary particle physics flow together into a single unified theory.... And when we have discovered this theory, it will be part of a true description of reality' (Weinberg 1998: 52). In his argument Weinberg is articulating a deep desire for the solidity of knowledge structures, if not certainty, because as we have argued in chapter 1, uncertainty is inescapably part of the modern – or post-modern? – condition. There is a longing for secure knowledge that is relatively impervious to the influences of the 'soft' parts of knowledge that surrounds its 'hard' core. Weinberg's argument is reminiscent of the distinction which Yehuda Elkana has made between images of knowledge and the body of knowledge (Elkana 1981). Images of knowledge make up the 'soft' part; the body of knowledge its 'hard' core. But this alignment understates their significance; 'soft' is not insignificant, subordinate or secondary. These images are socially constructed views on knowledge; they define for each culture, society, group or community the sources of knowledge, methods of its production, standards of proof and means of legitimating knowledge. Images of science also determine the relationship between knowledge and prevailing norms, values and ideologies, its locations and audiences.

As such these images are of great importance; they heavily influence, positively or negatively, problem choices in the body of knowledge, its 'hard' core. The images of science, its 'soft' part, build essential bridges between societal values and the 'hard' body of knowledge. For example, because of the growing professionalization of disciplines, an image of knowledge which gives a high priority to competence may come to dominate other images of knowledge which emphasize subject content, methodology or epistemology. The increasing emphasis on vocationalism in higher education and instrumentalism in research may reflect just such a shift between competing images of knowledge. Nor is the demarcation between 'soft' and 'hard' knowledge, between images of knowledge and the body of knowledge always clear; their interaction is also producing greater volatility. Yet, in specific social, cultural and scientific environments, a distinction can usually be drawn between the images and the body of knowledge. The images determine what is considered important, interesting, risky, symmetrical, beautiful, absurd or harmonious. None of these evaluations can be determined within the body of knowledge itself (Elkana 1981). But the majority of philosophers of science, whether Carnap, Popper, Lakatos, Quine or Putnam, and whatever their differences, refuses to consider epistemology in such sociological terms. So does a majority of scientific practitioners. For both it goes against the grain.

What is the reason for their refusal? Weinberg is defending not simply the now defunct super-conducting super-collider, the first of the large projects to be cancelled by the United States Congress; he is defending, far more fundamentally, what he believes to be an essential element of the scientific enterprise, a certain notion of Reality and Truth (even if we accept his own caveat 'whatever that is'). But how widely can such a vision of a 'hard core' of scientific knowledge impervious to social and cultural contextualization be applied across the wide spectrum of heterogeneous theories and practices now described as 'science'? How many other scientific disciplines, or sub-disciplines, have matured to the point where, in Weinberg's terms, they are able to produce 'permanent accomplishments'? Who is to determine, and how is it to be determined, what that point is and when it has been reached – exclusive scientific communities (which, in any case, are increasingly problematical to categorize) or a more open process of contextualization? How many scientific fields can satisfy the test proposed by Weinberg: 'it is now considered to be much more acceptable to base a physical theory on some principle of "invariance" (a principle that says that the laws of nature appear the same from certain different points of view)' (Weinberg 1998: 50)? Can this test be applied to sprawling phenomena such as magnetism or superconductivity, which are part of much more complex and unruly systems (in contrast to the highly simplified and formalized systems which characterize high energy physics)? This accumulation of difficult questions tends to undermine Weinberg's confident assertion that a *cordon sanitaire* can be maintained between 'soft' and 'hard' science, the images of knowledge and its body. Indeed, he provides a fine example of a shifting image of knowledge (as opposed to changes in the body of knowledge) when he writes: 'when a scientist reaches a new understanding of nature, he or she experiences an intense pleasure. These experiences over long periods have taught us how to judge what sort of scientific theory will provide the pleasure of understanding nature' (Weinberg 1998: 50).

A powerful reason for this reluctance on the part of scientists to acknowledge that socially constructed visions of science may fundamentally shape, in Elkana's phrase, the body of knowledge, its 'hard core', is that it appears to constrain the novelty for which science constantly strives. The idea of scientific revolutions successively sweeping away old theories and installing new ones is very attractive; its quasi-Darwinian rigour gives the scientist a privileged role as the secular priest of modernity. Popular visions of science, and the social interventions such visions may provoke, are seen as potentially restricting the dynamism of science. This helps to explain the ambivalent reception of Thomas Kuhn's work. On the one hand his notion of periodic scientific revolutions leading to fundamental paradigm shifts is appealing because it appears to reinforce this impression of a dynamic science steered by discrete scientific communities; even the intervening periods of 'normal science' reinforce the idea of a self-sustaining and self-referential scientific system. On the other hand Kuhn's thesis of incommensurability seems to undermine that idea by emphazing the historical contingency and cultural contextualization of scientific revolutions. Paradoxically it may even be during periods of 'normal science' that science is most autonomous, free to add 'permanent accomplishments' to the body of knowledge, while it is during periods of scientific revolution that visions of science are most potent – and either constraining or liberating depending on circumstances.

As has already been pointed out, Weinberg criticizes Kuhn for not distinguishing between change and continuity in a scientific revolution. According to David Mermin, many social scientists working in science studies fail to appreciate, and therefore to explore systematically, what he calls the exhilaratingly narrow line between radical and conservative aspects of introducing new scientific points of view and trying to understand their implications. Valid features of old points of view endure and often continue to be recognizable within the new framework. Newtonian mechanics are alive and well, and successfully sending mobile cameras to Mars, despite their 'overthrow' in both the relativity and the quantum revolutions. To a large extent Kuhn was responsible for the lack of attention paid to the conservative aspects of scientific revolutions. It is a radical experience to take something which apparently is well understood and look at it from an entirely different perspective, because this experience has to be re-thought in ways which are foreign, even incomprehensible, in the old perspective. But it is a conservative experience because the new perspective must preserve, with negligible alteration, large amounts of old and well-established knowledge – even if the language in which this knowledge is expressed has to be drastically modified (Mermin 1998: 605–6).

There appears to be a fundamental difference of values and feelings – on one side, the search for the historically invariant, sometimes expressed in absolute terms as the search for invariable and eternal truth that offers direct and unmediated access to reality; on the other, the pleasure and playfulness that come from new and unfamiliar discoveries, combined with the thrill of novel perspectives and radical insights. But, of course, it is misleading to present these differences dialectically because both play their sides of scientific sentiment in the praxis of science. This is why the deliberate restraint shown by scientists in evoking the public image of science is both puzzling and revealing. It is as if all traces of the subjective and the personal – motivations, delights and frustrations – must be totally erased; they are censored from the official accounts of science. Only objective and impersonal aspects of a self-propelling and a self-referential science, the advance of an untainted reason, are permitted. Perhaps the principal purpose of the epistemological core is to act as a shield to protect science against potential attacks and ignorant criticism and to preserve the vulnerable social space of institutional and cognitive autonomy of those who work within its boundaries. Even language is considered dangerous and slippery; only mathematical formulae and equations can be trusted to withstand the onslaught of change.

But when scientists-as-people are allowed to enter the picture, the balance between 'hard' and 'soft' parts of science, between the vision of science and the body of knowledge, shifts. The core shrinks to a small number of irrefutable and invariant laws, which resist contextualization within a wider social environment. The existence of these natural laws is not contested – even in the sharpest public controversies about science and its uses/misuses. If there still is an epistemological ground, it consists of different clusters of local and heterogeneous practices. In different ways and to varying degrees these clusters incorporate the materiality of scientific activities and their links to the concrete contexts in which they are embedded. There are shifting interactions between images of knowledge, the criteria which are used to assess the validity of scientific theories and the utility of scientific practices on the one hand, and on the other the body of knowledge. These judgements are made by people who communicate with other people. In doing so they construct narratives which bind them together into scientific communities and lend coherence to their projects and activities. It can even be argued that scientific creativity is enhanced when attention is paid to the 'soft', the social, even the subjective. Visions of science are not distractions from, or distortions of, the body of knowledge; rather, visioning is often antecedent to the production of knowledge.

What emerges is a more nuanced, more sociologically sensitive, epistemology – which is more resilient than the 'hard core' of autonomous self-referential epistemology which scientists have struggled to articulate and to defend. If science is to engage more strongly with the *agora*, its image of impersonal, 'objective', self-organizing structures purged of allegedly subjective elements will need to be complemented and corrected by putting people, human agents, back into science. Our conception of science has to find room for the wide range of people who engage in material scientific activities and are linked in concrete ways to other social spaces in the *agora* that go far beyond the laboratory. Rather than trying to protect a 'hard core' which turns out to be empty or irrelevant for practical purposes, the many soft layers and clusters need to be strengthened by making their knowledge claims more socially robust. Putting the people back into science is not a superficial task to be approached cynically or opportunistically, a public relations exercise to make them more visible (which is often how the need to improve the public understanding of science is presented). Nor is the challenge simply to create a new image of scientists as more approachable, communicative or simply more humane – however important it is that scientists should display these qualities (which should be part of their training). At issue is much more than better public understanding of science and better education for scientists. Life in the *agora* will be more challenging, requiring a radical re-thinking of science itself.

In the context of what we have called Mode-2 society and of increasingly contextualized science, conventional notions of reliable knowledge need to be complemented by broader conceptions of socially robust knowledge – just as, at an earlier time, the quest for scientific truth was succeeded by the search for reliable knowledge (a process which replicated the even earlier shift from absolute faith, as revealed through religion or established by ageless custom, to scientific truth, based on inquiry and experiment). In other words, what is happening is not some sudden and jarring paradigm-shift – from science to non-science, or from universal standards of objectivity to locally determined conditions of relativism – but the latest stage in an evolutionary process of adjustment to an increasingly complex reality. Part of that process, of course, has consisted of a problematization of reality.

In the past half-century, science, despite its triumphant advance (or, perhaps, because of it), has claimed only a circumscribed authority. The argument has been not that science can reveal or recognize a higher truth (or even reality). The very dynamism of the scientific system has emphasized the provisionality of scientific 'results'; it has been fuelled by uncertainty. Instead the argument has been that the autonomy of scientific institutions is justified because it creates the conditions under which reliable knowledge can be produced. It enables creative hypotheses to be proposed, validated (or disproved) through a rigorous process of inquiry and experiment, tested against alternative hypotheses within scientific communities which are both open and expert, and finally discarded or refined. The case for science, therefore, is instrumental, pragmatic and functionalist; in a phrase, it works. As a result, scientists have been able to ignore, or take for granted, the epistemological foundations of their authority. They have also been able to acknowledge that objectivity may be highly complex, problematical, even contingent without this acknowledgement undermining to any significant extent their day-to-day confidence in the validity of their work.

Today science apparently no longer 'works' as well as it did in the past – although it is more accurate to say that science now 'works' too well, with the result that there has been an explosion of expectations and, consequently, an accumulation of contradictions and contestations. One reaction has been a further retreat, an even more total abandonment of epistemology. The 'relevance' of research is all. A second has been a retreat to the high ground of a resurrected epistemology, to protect science from the forces of both relevance and irrationality. We believe there is a third way. A more nuanced and more sociologically sensitive epistemology is needed which incorporates the 'soft' individual, social and cultural visions of science as well as the 'hard' body of its knowledge. Within the wider environment in which science will have to work in the future, which we have called the *agora*, a disembedded and self-organizing science striving to discover invariant rules and accumulate knowledge will need to be complemented, if not replaced, by a new vision of science, richly contextualized, socially robust and epistemologically eclectic.

Conclusion

In the last two chapters we have developed three main arguments. The first has been that the basic conditions which have underpinned the production of reliable knowledge – described by John Ziman in terms of consensuality and consensibility – are not necessarily compromised by the shift to socially robust knowledge. To the extent that they have worked in the past – which, of course, has varied according to time, place and discipline – they can continue to be effective. Indeed it can be argued that, if consensuality and consensibility have been undermined, it has been as much by the reductionism of scientific practice as by any attempt to widen the range of stake-holders or more systematically to articulate the context within which science is produced.

The second argument has been that reliable knowledge has always only been reliable within boundaries, in two senses: first, it was inherently incomplete because science, above all, is a method; and second, to achieve reasonable reliability the problem terrain (and, therefore, the peer group) had to be restricted. Both are mirrored in the shift to socially robust knowledge: first, knowledge is incomplete – no longer only in the conventional sense that it will eventually be superseded by superior science (for Popperians) or by a new scientific paradigm (for Kuhnians), but also in the sense that it is sharply contested (and no longer within the controlled environment of scientific peers but in the wider *agora*); and second, this shift involves re-negotiating and re-interpreting these boundaries. As science moves into the wider *agora*, the subject of the next chapter, the boundaries within which reliable knowledge was once contained are dramatically extended. Contextualized science cannot be validated as reliable by conventional discipline-bound norms; while remaining reliable, it must be sensitive to a much wider range of 'social' implications (a much better word than applications, because no worthwhile knowledge is without implications, however diffuse they may be in spatial terms or currently unknowable in a temporal sense).

The third argument is that the epistemological core is empty – or, more accurately, that the epistemological core is crowded with many different norms and practices which cannot readily be reduced to generic methodologies or, more broadly, privileged cultures of scientific inquiry. As has already been stated, the case for science has often been made in essentially functionalist terms. It is – merely – a method for advancing knowledge; the theory, or conceptual foundations, of that knowledge is less important than its substance. '"What is truth", said jesting Pilate' – Bacon's famous introduction to his essay on truth sums up the view taken by many scientists. As long as science 'works', arguments about objectivity are unimportant. It is perfectly possible, therefore, to combine relativistic epistemology (if only by implication or through ignorance) with positivistic practice. Indeed it can even be argued that too searching an inquiry into the nature of knowledge, or of its epistemological core, is likely to constrain the potential of science; too precise an epistemology, like too rigid a methodology, could be taken to imply that ultimate and absolute truth is still attainable. The evident decline in the normative-cognitive authority of science did not seem to matter in the face of its incontestable and increasing utility. Its techno-authority remained unchallenged.

13

Science Moves into the Agora

In chapters 11 and 12 we discussed the content and condition of the so-called epistemological core – that set of inviolable principles, rules, methods and practices which are said to constitute the essence of science and cannot be discarded without endangering the whole enterprise. Next, we considered the movement of contextualized knowledge into its context of implication, a far more radical manoeuvre than the context of application. As a result of this movement, reliable knowledge, traditionally (and presently?) considered to be the hallmark of science, has been superseded by a richer and more resilient form of knowledge – which we have called socially robust knowledge. In this chapter we develop the concept of the *agora*, the social space in which this transformation takes place. In the eyes of many scientists this space is notable, even notorious, for the apparently interminable series of science-based controversies and, more threatening still, public contestation of science. Next, the place of institutions in the *agora* will be considered. Even if their anchorage is less safe and their contours less clear-cut than under conditions of modernity, institutions still play an important role in this social space where science is formed. The *agora* is not an unstructured and formless post-modern space.

Niklas Luhmann's provocative verdict is that 'society is shocked by its risks because there is no solution to this problem. The problem has structural sources and it is reproduced over time' (Luhmann 1996: 18). Luhmann, of course, does not deny the need to assess, manage and contain risks – in particular, risks arising from out-of-control technologies and wider uncertainties about the balance between potential benefits and damage when research embarks on new and untried directions. Rather, what he attempts to show is the pervasiveness of risks of all possible kinds in modern societies, linked to financial investments, career and marriage choices, life-style preferences. In this sea of uncertainties scientific and technological risks constitute only a small part – even if their consequences can be devastating. Since risk is inherent in all acts of decision-making and in modern societies decision-making is accorded a historically unprecedented importance, risk is pervasive. 'The problem will not go away'; there can be no safe way of making decisions. Instead decisions are events we use to distinguish between our observations of the past and anticipations of the future. Many aspects of life once taken for granted, whether as continuity or change, are now matters of decision. The future, of course, is more uncertain than ever because it depends upon the decisions of many others. The complexity – and uncertainty – are increased when, in the language of systems theory, risky decisions of one system (or individual) become a danger for another system (or individual). As a result both systems attribute risks and dangers to themselves and to others. This is what is meant by 'society being shocked by its risks', in Luhmann's phrase. The problem of risk becomes a problem of attribution, internal to modern society. Consensus or dissensus in attributing causes and effects replace consensus about an 'outer reality'. All kinds of 'reasonable standards' and other normative criteria (although they continue to be important in other respects) become dissolved through the attributive mechanism of decisions which is entirely internal to society.

In recent decades the number of risks has proliferated in the borderland between science and technology on the one hand and society on the other. A curious kind of reversal has occurred. Science (and to some extent the technological mastery of nature and technical artefacts) has been regarded by the public as a major source of uncertainty and risk, especially with regard to environmental hazards. But science had been able to cope with its own science-induced uncertainties. Today many scientists find it more difficult to cope with these uncertainties; they have become haunted by public fears about scientific research. Not only do they regard these fears, expressed in a series of public controversies about both the probability and the acceptability of specified risks, as unpredictable and often misplaced; they see these fears as a source of new uncertainties. Risk feeds off risk. The creation of ethics committees, the development of ethical guidelines and an apparently endless stream of regulations, procedures and protocols, checks and controls have not only disturbed the routines of laboratory life and rhythms of grant-getting; they have also made many scientists aware, perhaps for the first time, that the public sees their work in a different (and more critical) light. The interface of exchanges with the public has thus received a new slant – that of contestation.

Earlier in this book we drew attention to the historical emergence of a public space neither controlled by the ruler nor relegated to the private sphere. With the triumph of free-market capitalism and liberal democracy, this public space was transformed into an arena not only for market exchanges but also for open political discussion, an arena where criticism could be voiced openly, where public opinion was formed, and political consensus reached. Increasingly, the desires of both 'consumers' and 'citizens' were articulated in this public space, in the different forms identified by Albert Hirschman – exit, voice and loyalty (Hirschman 1970). As Mode-2 society and Mode-2 knowledge production co-evolved and co-mingled, the deeply rooted instinct of scientists to shy away from politics was no longer effective. Not only were the boundaries of science transgressed by controversies and contestations, scientists were no longer protected (and privileged) by their disavowal of politics. Although only a few scientists would accept Latour's vision of 'science freed from the politics of doing away with politics', many view the emerging *agora* with suspicion (Latour 1997).

The *Agora* Revisited

We have chosen to use the term *agora* to describe the new public space where science and society, the market and politics, co-mingle, because of its association with the original *agora* in the city-states of ancient Greece and also because we needed a novel, and expansive, term for a space that transcends the categorizations of modernity. According to Plato, Socrates outspokenly espoused an outrageously elitist (in modern terms) position. He supported the aristocratic faction in the Greek *polis* which possesses and understands the immutable 'laws of geometry' against the vagaries, distortions and irrationality which characterize the formation of public opinion and the articulation of political will among the 'mob'. In an effort to overcome the corruption and disorder which, in Plato's view, threatened his native Athens, his solution was to find an eternal anchoring point outside the cacophony of political, social and economic interests. The lure of such an 'external' certainty – whether enshrined in the laws of the gods, of geometry or of nature – is still strong, even at the turn of the twenty-first century. The restlessness of the modern *agora*, like its ancient counterpart, challenges this classical vision of eternal truth.

The argument about who has (exclusive?) access to the realm of nature and the natural order and, therefore, who properly comprehends its laws is still at the root of tensions that persist to this very day. The recent 'science wars' were a painful reminder of attitudes first articulated in Plato's Athens two-and-a-half millennia ago. But some things have utterly changed. A review of public attitudes towards scientific controversies strikingly reveals that what is contentious and contested is not the laws of nature; nor is the debate a clash between elitist and populist principles about the definition of science. On the contrary, the modern *agora* has been moulded and shaped by the Enlightenment; only a few of that movement's more presumptuous (and un-scientific?) ideas have been thrown into the dustbin of history. Science had a recognized, even privileged, position in the Enlightenment. Consequently it was irreversibly linked to the project of modernity. Today it is a – perhaps the – key input into debate within the contemporary *agora*. Post-modern thought is merely an epiphenomena – unable to disinherit its predecessor, while groping to find ways reflexively to re-constitute, re-invent or to re-think it.

The contemporary *agora* can hardly be said to be populated by 'the mob'. It consists of a highly articulate, well-educated population, the product of an enlightened educational system. The forces of democratization have stimulated the growth of mass systems of education, at primary or elementary levels in the nineteenth century, at secondary level in the early and mid-twentieth century and at post-secondary and university level since 1945. According to the principles of the best educational tradition of the Enlightenment, graduates should have learned to be more rather than less critical, more rather than less reasonable. Experience of participating in liberal democracy should also have taught them how best to express their views and voice their demands. Moreover, the shift from a culture of the autonomy of science to a regime of greater accountability presupposes that science can – and ought to – respond to these demands. Greater attention to what people 'really' need and want presupposes the ability and the political will to act accordingly. Today there is a widespread expectation not only that science ought to listen to these demands but also that science can satisfy them. The incorporation of science into the *agora*, therefore, is an expression of confidence in its potentiality, not a loss of trust.

Other changes that arise in the context of Mode-2 society point in the same direction and reinforce, in a co-evolutionary mode, similar underlying tendencies. The process of individualization which has reached such a culminating, unique manifestation in Western industrialized societies has brought with it also a heightened awareness of the empowerment of the individual, even if it is also acknowledged that such empowerment operates under new, equally powerful, economic and social constraints. Today there is a historically unprecedented emphasis on agency – rather than on structure which was the hallmark of modernity – and also on the necessity for the individual to intend and to act, to believe and to act, to make choices while acting – and to be as fully aware and informed as possible of the intentions, beliefs, choices and actions of others. Instead of structure, we are faced with processes of structuration (Giddens 1984).

Individualization is inevitably accompanied by an upgrading of the subjective, including a revaluation of the subjective experience. In many fields apparently remote from the celebration of an emancipatory subjectivity, we can sense the spread of a different attitude and of taking seriously what had previously been discarded as being 'merely subjective'. This is in evidence, for instance, in the field of health care. Programmes for patients' education have been established as a device to allow doctors to learn what they do not yet know about their patients – including, for example, the actual behaviour of patients with regard to obeying (or disobeying) drug prescriptions. Once clinical success was appraised almost exclusively in terms of 'objective' (because quantifiable) measures such as mortality rates, physiological measures such as blood pressure or diagnostic test results that are surrogates for physiological functions; now such measures have been replaced, or at any rate complemented, by more subjective measures. Patients' perceptions of their own state of health, measures based on patients' preferences and other health-related quality-of-life aspects and subjective assessments are increasingly important because clinicians and patients now have to make decisions associated with different types of outcomes such as length of survival, preservation of function, or pain relief (Clancy and Eisenberg 1998: 245–6).

Such a shift to the value and place of subjective experience, preferences, perceptions and values demonstrates – not only in the field of health – how successful contextualization has been. The context in which knowledge is produced can be extremely demanding. But it also enriches, differentiates, expands and transforms this knowledge. There is an interesting parallel here with the 'customization' which has become so prominent in industrial production and in symbolic and material design. By taking more account of individuals or locally specified functions (treating them as customization does new materials) or by taking individual preferences more seriously (like software programmes or design specifications), customization can mean much more than the advertising travesty of 'your personal copy'. By paying attention to the unique features or the specific configuration found in a context of application – whether this is the individualized, genetically-based drug design of the future or more socially and subjectively comprehensive forms of health care or the customized, specified functions that new materials are supposed to fulfil – new models of the new scientific contexts produced by specific forms of contextualization are emerging.

In the modern *agora*, therefore, these demands and desires are negotiated and re-negotiated, producing ever more novel forms of the contextualization of the knowledge. The *agora* is populated by a diversity of individuals who combine the roles of 'citizen' and 'consumer', while at the institutional level typically markets and politics set the rules within which this ceaseless process of negotiation and re-negotiation takes place. Under conditions of modernity (or, in our terms, Mode-1 society) science was still seen as an external force, which was brought into the public space in highly selective ways; it appeared in different, even contradictory, guises and images and to some extent mimicked the social stratification prevalent in the larger society. In a Mode-2 society all this has changed. Science is now an internal force, pervasively (if still reluctantly) present in the *agora*. Of course, science has always had its interlocutors; Galileo interacted not only with his aristocratic patrons but also with the craftsmen of the arsenal in Venice. But in a Mode-2 society the nature and intensity of such interactions has become transformed. It is no longer characterized by the stable and predictable triangular relationship between state, science and industry which held sway in the decades after the Second World War – and in which physicists were especially prominent because of their singular contribution to victory. Now, the 'context' itself has multiplied beyond recognition; it has become fragmented, on the one hand localized and on the other globalized; it can no longer be contained, as we showed in chapter 1, within the categories of 'market', 'state' or even 'culture'.

Of course, it is ahistorical to argue that the relationship between science and its social and political interlocutors was ever unproblematic; even the relationship between science and the military was much more tormented and tempestuous than simplified retrospective accounts suggest. However, today the degree of complexity, unpredictability and volatility – but also of intensified mutual interdependence – has increased many times over. Science and scientists now face an *agora* with multiple publics and plural institutions, such as the mass media, which vigorously conduct their own negotiations. They are faced with a complex bureaucratic and administrative web of funding agencies which devise their own policy goals, guidelines, assessment procedures and allocation mechanisms. The science-policy landscape seeks to accommodate an increase in private financing while becoming much more discriminating and demanding in public funding. Researchers have had to develop a new range of skills in communicating with their potential funders, writing grant proposals and promising enticing (but plausible) outcomes which cannot really be specified in advance. They face an industry-business landscape which itself is the object of radical restructuring.

The former large state industries, once science's most reliable and familiar partners, are themselves under attack. They stand accused of being ineffective, in terms of stimulating innovation, and inefficient, in terms of increasing productivity. Small and medium-sized enterprises, although much vaunted by policy makers, often lack the human and financial resources, and infrastructure, to be attractive to universities as alternative partners, although this is changing rapidly, especially in areas of advanced technology. There are also important differences between research fields; some, like the biomedical sciences, have successfully embraced the privatization of research, while others are still struggling to find their feet in an altered science-policy and -funding landscape. However, the greatest challenge to science in the modern *agora* does not come from the market and its exponential demands; nor does it come from a restructured industry which seeks to reduce its traditional commitment to in-house research. It does not even come from the changes in the public funding of science or the emergence of new entanglements, different science-policy alliances with perhaps divergent criteria. Although these new alliances will make novel demands on researchers, these demands can be accommodated, because the overall level of funding for research is not declining.

The main challenge in the *agora* comes on a deeper level. It arises from contextualization itself, which has so greatly enhanced the success of science, opening up new research fronts, expanding research horizons and establishing new and apparently flourishing research fields. But contextualization has a price. Science and scientists have not been used to the context speaking back, so it is not surprising that they see contextualization as a challenge to their cognitive and social authority. Caught on the defensive, they blame contextualization, and the *agora* in which it emerges, for the rise of anti-science sentiments, for the subversive influence of social scientists and other 'relativists'. They fear that irrationality will break through the fragile crust of scientification. There is hard evidence that such fears are exaggerated; public opinion polls and survey data consistently find that the decline in overall trust in science is not worse than the decrease in the legitimacy of other public institutions, and that science is still highly regarded in its overall problem-solving capacity. But none of this contrary evidence is likely to convince sceptics that the *agora* is not a hostile environment for the production of good science. However, any attempt to sustain an anachronistic image of science, which does not recognize the new realities of the *agora* and the new demands articulated through contextualization, is likely to be a self-defeating strategy.

There are other more hopeful strategies. For example, public contestation almost never leads to demands to dispense with scientific objectivity. Rather the reverse; it intensifies the search for better science. The Cumbrian sheep farmers researched by Brian Wynne complained about the official scientific experts whom the government had sent to investigate the extent and consequences of radioactive damage caused by the Chernobyl fallout on their land, because their own local knowledge about the variation in the acidity of soils had not been taken into account. In insisting that this local knowledge should have been part of the overall 'scientific evidence' used to shape future policy, they were offering their own, arguably superior, account of 'scientific objectivity'. They were equally committed to 'methodological rigour'; their complaint was that the official scientists had failed the test miserably (Wynne 1996).

Genetically modified plants provide another example – in the form of a currently fashionable mediating exercise in an area of controversy. A quasi-experimental set-up was used. In Germany, experts in favour and experts against the licensing of GM plants met in a series of prolonged face-to-face interactions. The aim was to sort out as far as possible their underlying differences about how the problem should be framed and how to assess scientific evidence about the probabilities of risks. To their surprise, the investigators found that both experts and counter-experts used similar kinds of 'rational' argumentation. Both sides conformed to the canon and rules of a proper scientific discourse, weighing the available evidence and seeking to eliminate untenable or empirically doubtful hypotheses. The differences that persisted arose from different perceptions of risk and assumptions about how risks should be properly assessed. Often, as in the case of GM plants, conflicts about scientific facts turn out on closer examination to be inseparably tied to social and political values. Science has no authoritative answers to such questions. It cannot determine what kind of society people would like to inhabit or decide whether it is right to encourage, tolerate or restrain the development of aggressive agro-business (van den Daele 1996). At the end the tentative consensus that had been reached broke down when private face-to-face interaction ceased and the counter-experts had to face their constituency and the media as part of the public. The political realities of the *agora* proved stronger than an essentially academic encounter among peers. Another example can be cited from the Swiss referendum which took place in April 1998. The initiative was to outlaw bio-technical engineering and related research. Some of the most vociferous proponents of the referendum came from a relatively small circle of highly educated persons with a background in biology and, among them, women had an unusually high representation (Schatz 1998).

By taking seriously the reflexivity of the context, its capacity to 'talk back', scientists can discover the extent to which the contextualization of their research activities has worked. They may have developed hypotheses about whether, and if so in which ways, the subjective experience of people has been affected or about what other more long-term or large-scale societal changes are likely to have been produced by their research activities. Most researchers, through the peer-review process, are aware of the impact that their work has on their immediate colleagues; many also want to know whether their research extends to other specialities or disciplines; but few scientists have any real incentive to think about its effects upon the wider social context. Similarly those researchers who collaborate closely with industry are fully aware of how industrial management reacts and acts and, therefore, can interpret or even anticipate industry's attitudes towards and decisions about their research; but they do not have the same intimate awareness of the reactions of the wider public. But this is what the modern *agora* is all about: the development of such intimate and interactive, as well as anticipatory, awareness. The *agora* embraces much more than the market and much more than politics. As a public space it invites exchanges of all kinds, and creates a context in which wishes, desires, preferences and needs can be articulated as well as demands.

It is also a space in which particular forms of contestation are allowed, and even encouraged, at least in Western liberal democracies. Although the *agora* is a structured space, it is wrong to attempt to subdivide it again into sectors like markets, politics or media. In a Mode-2 society these forms of differentiation are beginning to break down, and to be replaced by fluid and dynamic (and pervasive) interlinkages. As a public space, the *agora* is shaped by the interaction of its actors/agents. Some are more visible, easier to identify and recognize and more powerful than others. But the *agora* is also a space in which different perspectives are brought together, ultimately creating different visions, values and options. In this sense context contains also contingency – but contingency that operates in this public arena within certain constraints. These constraints take various forms. The first is the processes of democratic debate, contestation, protest and negotiation. The second form is the economic constraints under which markets operate, increasingly in a global context. The third form of constraints is government policies, laws and regulations, which are far from being completely subservient to market interests. A fourth form, although of a different kind, is represented by the media and their influence on public opinion. But these distinctions between different forms of constraint merely reflect our inability to conceive of interlocking contingencies and constraints that arise in different parts of the *agora*, involving heterogeneous but partly overlapping actors and producing different, even divergent, outcomes. The grounds of the *agora* are shifting continuously, as are these interconnections.

The Place of Institutions in the *Agora*

We have already argued that the contextualization of knowledge production demands also a contextualization of the processes with which this knowledge is tested, verified and validated. In other words, the 'objectivity' and 'reliability' of contextualized knowledge demands a widening of procedures, methods and quality criteria to assess outcomes. As a result, our notion of scientific reliability must be extended in what we describe as the shift from reliable knowledge to socially robust knowledge. The increasing demand for participation in the *agora*, whether expressed directly by the public or indirectly, on the public's behalf, by articulating 'user needs' more clearly, is not only the result of democratization. It is also evidence of the success of science which – through increasing contextualization – now addresses the concerns of a highly heterogeneous public. Participation offers a form of appropriating knowledge that would otherwise appear remote and arcane. As a result the shift towards socially robust knowledge is sometimes described as a shift from a culture of scientific autonomy to a culture of scientific accountability.

But accountability is not enough by itself. Many of the public controversies provoked by the success of science have taken place in, or even created, hybrid spaces in which scientific debates have not previously taken place. For example, the concept of risk, originally defined in narrowly scientific and technical terms, has been transformed by controversies and debates in which experts, counter-experts and lay persons have all been engaged. As a result all have been caught up in a learning process. Today the 'subjective' side of risk cannot easily be separated from an 'objective' side. Risks have social as well as psychological dimensions – for instance, the extent to which risks are considered to be 'voluntary' or not. It is not simply that counter-experts and lay people have taken the idea of risk, manipulating and exploiting it for their own ends; it is that risk, as a scientific concept, has been refined and re-defined. The involvement of activists in AIDS research is an example of how the active participation – in this case of a highly educated group who quickly acquired sufficient medical, biological and statistically relevant knowledge – reshaped the statistical reasoning underlying clinical trials and produced significant changes in how these trials were conducted (Epstein 1997). So, far from being merely a nuisance or a necessary, but irksome, price to satisfy 'external' demands of greater accountability, contestation has become an 'internal' force, an inherent feature in the contextualization of scientific knowledge. This change is, almost certainly, irreversible. And, if appropriately addressed – by making knowledge more socially robust in different contexts and sites of contestation – it can considerably strengthen support for science in a Mode-2 society.

The *agora*, therefore, is not an empty or an anarchic place; nor has the state retreated leaving the vacuum to be filled by the unbridled forces of the market. Of course, it would be naive to claim the *agora* is devoid of power structures and power struggles. Power certainly matters. Money continues to matter. But, at the same time, the *agora* is a special kind of public space, where many of the elements of Mode-2 society come together in novel ways. Science is no longer outside, either as a cognitive or quasi-religious authority or as an autonomous entity with its special access to the reality of nature. In the past science provided a unique kind of legitimacy for the project of modernity, by holding a unique monopoly on the definition of nature and of reality, which for a long time went unchallenged. It opened up and sustained a future horizon of expectations embracing not only scientific and technological but also social progress. Science and technology helped to glue together the consensus between otherwise divergent social groups, by providing actual proof of improvement in living conditions, and thus helped to define a common outlook and a common future project.

Scientific objectivity and the reliable knowledge it engenders, therefore, were hallmarks of modernity. But just as the universality and presumed homogeneity of the nation-state turned out to be an illusion, despite the fact that nation-states have spread around the world, so science and technology have found it impossible to maintain their previous success in underpinning social consensus within liberal democracies. In the eyes of many citizens, science and technology are now equated with their results and products. They have come to be seen largely as commodities, access to which should be democratically regulated and the allotment of which should also be fairly distributed. Consequently, what should be produced and how it should be produced must be embraced within democratic decision-making, not least in the light of the potential risks that some research may carry. As a result, particular institutions have attained new prominence in the public *agora*. Good examples are the increase in legislation and regulation and the role played by the law with regard to settling public and private disputes about scientific and technical risk and determining health and safety issues.

The role and power of the media have also been vastly enhanced in the *agora*. In the changing relationship between science and the public, the media have a dominant role in shaping, and reshaping, public images of science. 'Selling science', in Dorothy Nelkin's phrase, has also tended to shift the emphasis away from the more triumphant and celebratory aspects of science and technology to their more sinister and risky aspects (Nelkin 1987). News about science and technology now appears regularly – and prominently – on the business pages of newspapers, and science's capacity for innovation receives ever greater emphasis. But there have also been significant changes in the ways scientists relate to the media. In some prominent cases of scientific breakthroughs, a new phenomenon, 'science by press conference' or, in other words, the announcement of scientific findings to the press before the more tedious procedures of peer review have been completed, has demonstrated how successful the media have been in distorting traditional norms of scientific behaviour. In other cases scientists have used the media to create a stage on which they can present themselves, as personalities, and their ideas in ways which would have been impossible to achieve by traditional scientific means.

The *agora* has also become the arena in which social movements have first articulated their criticisms of science, in particular their demands for different priorities or for 'alternative research'. For example, powerful critiques about the objectivity and neutrality of science were developed and persuasive claims about a specific feminist perspective were made by successive waves of the feminist movement in the space provided by the *agora*. As a result, gender studies is now firmly entrenched in university curricula. Also, the underrepresentation of women in senior scientific positions has become an enduring concern not only for policy-makers, but also for professional societies such as the American Physical Society, which has tried to promote women scientists as role models. The ways in which clinical tests had been conducted had to be revised after disturbing research evidence about the underrepresentation of women in these trials and the scientific distortions produced by this underrepresentation. Self-help groups, often inspired by women's health groups, have not only made their voices heard, but have also been able to initiate research on otherwise neglected problems. As a result of such activities and the activities of other groups campaigning for patients' rights, the professional authority of doctors has been contested and partly has had to be redefined. Ethical committees and committees with patient representatives are now a routine feature of how hospitals are run, how clinical tests are screened and which research grants involving experiments with human subjects can be approved. The requirement to take account of the gender dimension has now become routine in many parts of the research system.

The environmental movement has also had a profound and permanent impact not only on public opinion but also on research practice. It has stimulated the development of a flourishing interdisciplinary field in the shape of environmental sciences, which continues to produce new alliances (for example, between economics and ecology). Ever since the degradation of the natural environment and the need to develop long-term strategies for sustainability were incorporated into the public research agenda in the early 1970s, environmental research has enjoyed high levels of funding and unprecedented public and political attention. Moreover, its prominence has contributed to the maturation of its scientific content, and the enhancement of its modelling capabilities and methodologies as well as a recognition of their limits. Although more strongly linked to economic incentives and governmental regulations, the related 'greening' of industry has also stimulated the growth of environmental research and its application. At times the learning processes involved in dealing with new policy issues, with their protracted intermingling of scientific and policy-related dimensions, have been difficult, as the many meetings preceding and following the major events of international agreements under the auspices of the United Nations or similar bodies demonstrate. Nevertheless, there have been major advances in how to deal with environmental issues on a global as well as on a national and local scale.

The role that non-governmental organizations (NGOs) have played in connecting knowledge to action in the environmental field, and linking 'local' knowledge to 'scientific' knowledge, can hardly be overemphasized (Jasanoff 1997). The medium- and long-term effect of social movements underlines how not only research agendas but also the substance of scientific inquiry can be reshaped by issues originally raised outside the research system. They also draw attention to analogous changes in Mode-2 society without which this reshaping could not have taken place. These changes include the 'empowerment' of individuals, the enhancement of their roles as citizens and consumers brought about through novel forms of collective action and interest formation, and their respective influence on the established political process, including the role and function of political parties. These processes, including a greater variety of protest movements, have significantly shaped the *agora* in its present form.

Although there has been mutual accommodation between old and new structures and forms of political articulation of interests, social movements and their agendas have become an integral part of the institutional make-up of the *agora*. In the next chapter we will consider the important part played by shifting conceptions of expertise – which we argue, like knowledge, is now socially distributed – in stimulating knowledge production in the *agora*.

14

Socially Distributed Expertise

In this chapter the role of expertise, still among the most powerful mediators between 'science' and its neighbours in the *agora*, will be discussed (Nowotny 2000). Expertise is at once contested, problematical, central and indispensable. It is the mechanism by which problems are framed, problems which often incorporate not only scientific judgements but also more basic, and deeper, social, political or cultural predispositions and commitments. In the Mode-2 society which has been outlined in the first chapter of this book, expertise, too, is becoming socially distributed. The emergence of the individual as decision-maker is one aspect of this wider distribution of expertise. In the final section of this chapter the implications of socially distributed expertise for the authority and image of science are considered. Can they be re-constituted – and, if so, how? Or is the gap between the image of science and its actual practice now too wide? Authority, like expertise and knowledge, is now more widely diffused through society. The ability to reconstruct the image of science may hold the key to understanding science in a Mode-2 society.

Modern society is – was – a society of experts rooted in a division of technical, professional and scientific labour. In the Mode-2 society which is emerging, expertise will continue to play a key role – but in different ways. In the last chapter we discussed the development of a new kind of public space, which went beyond the market and politics (and which we labelled the *agora*). Scientific expertise is one of the crucial mediating mechanisms in this new *agora*. Its wide-ranging authority is derived from scientific and technical knowledge, skills and know-how. But expertise reaches out beyond knowledge production by linking knowledge to decision-making and action. Although 'official' expertise is restricted to certified experts in some regulatory regimes (medicine and the law are the standard examples), expertise in practice is widely distributed. It is inherent in many everyday decisions. But the degree to which expertise relies on an explicit appeal to 'best practice' in a professional sense or to current standards of scientific knowledge varies. The content and form of expertise are context-dependent.

However, expertise transcends, as well as being dependent on, particular contexts. Typically experts are summoned to respond to specific problems or issues. But they must draw on a generalized and broad knowledge base. Narrative becomes one of the central ways in which the voices of experts are orchestrated to help produce a more wide-ranging epistemic, social, political or legal authority, which then is re-introduced to and feeds back into the specific context in which expertise is required. It is in this sense that expertise is both context-dependent and context-transcendent. Expertise claims to possess not only knowledge about a variety of concrete practices and their effects, but also knowledge about the links between the practices of the past and those of the (uncertain) future. In this way it is the key link between knowing and acting, between scientific and technological best practice and political decision-making. Potentially therefore expertise is as all-embracing as the practices to which it relates. It is crucial in a present which always seeks to command action with the urgency of the immediate. But the ways in which the narratives of expertise are enacted, and consequently the audiences which they seek to persuade, change over time. So it may be useful briefly to retrace the historical trajectory between expertise and decision-making in order to understand better the situation in which we find ourselves in a Mode-2 society.

Expertise, Decision-Making and Modernity

Despite its association with a culture of rationality, modernity was not seen as resulting from rational decision-making processes. Rather, the large-scale social and economic transformations that transformed the lives of individuals, communities and societies were initially regarded as quasi-natural forces, unpredictable in their origins, processes and outcomes. The 'hidden hand' of modernization could not be guided. Industrialization and urbanization and the uprooting of collective and individual identities which inevitably accompanied them were interpreted by contemporaries to be the work of impersonal forces which could not be channelled or resisted. The only recourses appeared to be resignation or exploitation, accommodation to inevitable suffering or entrepreneurial opportunism. Later when the tangible benefits of modernization had become more visible, with railways and electricity reaching out to remote rural villages, these beneficial effects were still not attributed to rational decision-making. Rather, the development of new technologies and modern infrastructures, and the consequent social changes, were seen as manifestations of the forces of progress. These technologies and infrastructures appeared to be derived from science's ever increasing capacity for innovation and utility. Their distribution was shaped by prevailing social hierarchies of class, status and economic power, and their development went hand in hand with the relentless expansion of the nation-state and its centralizing imperatives. Of course, there was resistance to the growth of industrial urban society (and its inevitable discontents articulated by a sophisticated intellectual elite); countercultures were a hallmark of modernity from the start. But there were few complaints about inadequate decision-making or the failure to create a consensus about scientific priorities or, in contemporary language, technological foresight.

Science and technology were largely absolved of responsibility for the distortions and dislocations of progress because they were seen as lying outside society. Eventually there were political interventions to correct or compensate for the least acceptable consequences of industrialization, which produced the welfare state and its systems of social security. But it is remarkable how deferential general attitudes towards the authority of science were (Felt 1997). Science itself was not to blame. Technology too was exonerated because it was seen as applied science or, at any rate, as derived from science. Instead both were celebrated; the beauty and grandeur of their achievements were admired. Attitudes to further progress and technological potential were often Utopian. Although the ambivalence of scientific and technological progress could already be sensed, the negative implications were downplayed. No traces of our present predicaments about the role of scientific and technical expertise in policy-making were evident. Any clouds on the horizon were smaller than a man's hand – if they were visible at all. Public controversies about science or technology were absent, except in narrowly prescribed cases. There were no attempts to reach international agreements in matters that have since become global environmental concerns. Modernity, it seemed, was still secure in its own foundations. It rested on the sense of control that it had achieved within the confines of the nation-state and within science and technology's contribution to making visions of future material and social progress come true.

But neither modernism nor modernity was effective by itself. Constant repairs were needed which emphasized its limitations. The processes that produced, and sustained, its effectiveness also had to be explicitly recognized (Deuten, Rip and Jelsma 1997). So modernity was never devoid of decision-makers and decision-making structures. Indeed it is possible to argue that only with the advance of modernity did decision-making become as potentially significant and pervasive as it is today. Modernity celebrated a culture of rationality; it is the centrality of decision-making that distinguishes modern society from early forms of society resigned to an inexorable fate. Decision-making already relied on scientific, technical and administrative expertise. This comprised a vast array of competences and skills, many of which served the expansion and growth of the nation-state. Nation-states, of course, competed with each other, sometimes through warfare and at other times in the economic sphere. Their success was measured not only by victory on the battlefield but also by the construction of railways (the two, of course, were intimately connected in nineteenth-century Europe). Railways and other industrial systems required not only the most advanced technology but also large-scale financial investment, often provided by foreign capital. The achievements, and failures, of material progress had important social reverberations. For example, bankruptcies, and boom-and-bust cycles, inspired Durkheim's analysis of social *anomie*. But strategic calculations had to be made and risks taken in other areas. The rise of the welfare state with its extensive social security systems depended heavily on the expertise of statisticians and actuaries; it was in this context that probability theory came to be widely used. The building of bridges and high-rise housing and the widespread use of machinery in manufacturing led to a strengthening of safety regulations and other controls. There was a proliferation of international conventions to standardize measurements. Similar efforts at standardization were made in national contexts as states elaborated their regulatory and administrative regimes.

This inexorable extension of the nation-state's core activities meant that new areas became subject to direct state control, and an ever denser network of administrative regulations was created. This was made possible by the expansion and professionalization of the civil service and the mobilization of other quasi-state experts. The growth of state power rested on their competence and expertise. As a result, higher education was restructured to satisfy the increasing demand for this competence and expertise. This led to a broadening of the base of the traditional disciplines in which civil servants had been educated. The *grandes écoles* established in France in the late-eighteenth and nineteenth centuries were designed to develop a unique blend of scientific knowledge and technical expertise in the service of the state, a pattern which was followed with variations throughout continental Europe. However, expertise took two forms; on the one hand, the cult of impersonal bureaucracy so insightfully analysed by Max Weber, and, on the other, the exercise of personal expert judgement. The tension between these two forms of expertise persisted. On the one hand there was, in Theodore Porter's phrase, 'trust in numbers' (Porter 1995) and commitment to the ideal of the impersonal bureaucratic expert; on the other hand the ideal of the independent professional practitioner, arrayed in liberal professions, commanded similar support. However, in both cases expertise was ultimately defined by its relationship to the state. The relative autonomy enjoyed by professions in some countries was balanced by their closer integration into the state bureaucracy in others. In both cases experts were the nation-state's agents of modernization, actively engaged in shaping the practice of its future and rooted in the grand narrative of modernity.

Of course, neither scientific and technical expertise nor the know-how of social engineering emerged spontaneously from their own diverse fields of practice. Both were carefully nourished, certified, controlled and transmitted through highly regulated hierarchies and networks. These networks linked institutions of higher learning, professional societies and corporate associations with the state's regulatory, legal and institutional frameworks. In nearly every case the state retained its effective monopoly of certification and, consequently, of quality control. However privileged the relative autonomy enjoyed by some professions, elaborate rules were devised to regulate access to professions and the content of professional training and to specify the conditions under which expertise could be exercised. These rules were a cage, or a cocoon, within which professional expertise could develop. These experts, whether members of liberal professions or state bureaucrats, were central to the process of modernization. That process, and their contribution, had three aspects. First, the overall goals and questions to be addressed were framed by the projects of modernization. Second, models of best practice and expert judgement were provided by the professions. But there was a third, less obvious, aspect which enhanced the status of scientific and technical expertise.

This third element was the knowledge base as it developed inside the universities in the latter half of the nineteenth century, namely the predominantly disciplinary structure of knowledge production which we have described as Mode-1 science. The demarcation between pure and applied science greatly enhanced the status of science, even in areas of professional expertise apparently remote from Mode-1 science. To appreciate the authority wielded by science at the height of modernity – an authority grounded in and legitimated by its epistemological status – it is important to recognize the social space occupied by science. Science and technology had become major allies of the nation-state's modernization projects; they helped to expand, and then to consolidate, the role of the nation-state. Accountability was exercised within the framework of the accountability of professional experts; Mode-1 science was outside this framework and, consequently, not accountable. Scientific and technical expertise were only able to share the incontestable authority of science to the extent that it could be demonstrated that scientific principles had been correctly applied. When expertise was compromised, it was because science had been wrongly applied. But the authority of science remained intact and incontestable.

The Rise of the Individual as Rational Decision-Maker

The contrast with late modernity could hardly be greater. The role of scientific arguments in structuring everyday social choices such as eating habits, medical treatment and life-styles; the use of technical evidence and its impact on trust and social cohesion; and frequent arguments about emotive and irrational public responses to issues presented in scientific or technical terms are all pervasive phenomena. But the vast increase of scientific and technological knowledge and the possibilities this increase creates for radically new forms of intervention, control and manipulation of nature are only half of the story. The other half is the rise of the individual as decision-maker, now faced with a bewildering and proliferating array of options. Decision-making has become part of his or her existence. The individual today is defined through the sequence of decisions which, in their contingency, make up his or her identity. The inexorable ascent of the individual, who is presumed to be a rational decision-maker, is now central to the functioning of the market. Not only is the primacy of the individual-as-decision-maker validated through market mechanisms; it is also used to vindicate the efficiency of the market. This is true of the individual's dual roles – as an informed and an information-seeking consumer and citizen.

The privileged status bestowed on the rational individual as decision-maker has implications for the ways in which collective decisions are reached, how they are implemented and what kind of legitimacy they enjoy. This is one of the main reasons why scientific and technical expertise has become distributed throughout society and also why, in the course of this diffusion, expertise has been to some extent privatized. Decision-making is no longer seen as densely compacted into the hierarchical structure of state bureaucracy or analogous administrative and business units. It has become disembedded from the centralized structural matrix of modernity where it was previously located. If, in contrast, late modernity locates decision-making in the desires and demands of the individual who – in principle, at any rate – has access to all the information required, expertise also becomes a highly individualized commodity. In the fluidity of information exchanges and the proliferation of options, expertise takes on a new guise. It becomes a fluid configuration of knowledge, information and experience, all of which are most likely to resonate with questions arising in highly specific and localized contexts. As a result these individualized and decentralized forms of expertise have triumphed over the hierarchical and centralized forms that characterized modernity. We are all experts now.

Of course, this new kind of diffused expertise, linked to a decentralized and individualized decision-making structure, cannot replace (yet) the culture of public expertise, backed by professional standards and scientific authority. But this new expertise potentially gains collective strength because it is distributed throughout society and, consequently, can build powerful networks with other sites of expertise. When the private expertise of many individuals has become public, their combination may come to form a kind of public expertise. Typically, this transition from private to public expertise is always mediated – through the media and public opinion, through various organized bodies and committees, through social movements or self-help groups, through political processes that seek wider public participation, through discussion groups on the Internet or more traditional channels of political association, lobbying or interest articulation. All these resources can be mobilized to tap the widely dispersed private expertise of individuals in order to provide the raw material for public expertise to emerge in contested areas and issues.

This helps to explain the paradox of scientific and technical expertise, the dis-equilibrium between public and private expertise. More expertise is needed and demanded; more expertise is being produced by the myriad of individual decision-makers. But publicly available and certified expertise inevitably lags behind because public expertise does not, and cannot, equal the sum of all private expertise. Just as nobody can participate in all decisions that affect him or her, expertise that is both publicly available and customized for individuals cannot be generated. Expectations remain unfulfilled, and the links to private experience unpredictable, volatile, fragmented and diverse. Expertise has become disembedded in the process of liberating itself from its previous ties that bound it to the public realm and to the centralized decision-making structures of modernity. Hence the growth of discontent in the interstices between private and public arenas. The more that public expertise is expected to mirror the individual's private experience and individualized expertise, the more disappointing it must become.

For some this may appear an exaggerated account of the transformation expertise has undergone in the shift from modernity to late modernity. But it is important to highlight the role played in this transformation by the emergence of the individual, rational and informed decision-maker. This character has come to play a central role in today's narrative of expertise. The more numerous the options, the more information is needed. But, at the same time, criteria are being sought to reduce the complexity of this information in order to make it manageable. The market and politics both offer solutions to this dilemma between the proliferation of choices and the explosion of information by providing individuals with many entry points for further contestation and mobilization, for action and appeal. The monolithic decision-making structures of modernity have been demolished and replaced by individual participation and choice. As a result, the extent to which liberal Western democracies depended upon the continuous necessity of negotiating consensus, even in areas that previously were thought not to lend themselves to compromise, has been exposed. This is where the contestation of the public authority of science enters the argument.

Late Modernity and its Discontents

Expertise has always been pragmatic. It has also been defined in particular contexts. This pragmatism and contextuality present a major challenge for experts. They have to draw on a stock of knowledge available at the time when answers and action demand it. However, the substance of the narrative, the questions which their expertise is expected to answer, are slipping out of the grasp of experts. This substance, and these questions, are increasingly determined by stake-holders, not experts. It can even be argued that scientists, in their guise as experts, do not respond to questions which they have chosen – in contrast to research, where they set the questions. The questions experts are expected to answer as experts typically arise from the need to choose between particular options rather than produce generalized answers. Consequently they are often forced to transgress the limits of their competence (Roqueplo 1997: 36). So strategies have to be devised by experts in order to enlarge, mobilize and tap into their knowledge bases. Often experts are under instant and intense pressure to respond to a crisis in decision-making, the most recent example being the frantic efforts by the United Kingdom government and the European Commission to confront the questions raised by BSE. As a result scientific experts have to synthesize all available and relevant knowledge, interdisciplinary as well as disciplinary. Of necessity experts have to transgress the boundaries of their disciplinary knowledge and the constraints of their own limits of knowledge. There are obvious parallels to Mode-2 knowledge production, where transdisciplinarity is achieved by focusing on research problems as they emerge in contexts of application and where the heterogeneity of knowledge producers introduces additional criteria of assessment apart from scientific quality.

The demands for scientific and technical expertise in a socially distributed knowledge production system differ from the usual demands on research. Because experts have to answer questions chosen by others, they cannot adopt the traditional tactic of reformulating questions in such a way that they are amenable to scientific investigation. Experts have also to draw on additional resources. But the main difference is that experts – in contrast to researchers – must respond to the imperative of the 'immediate'. They often work to very tight deadlines imposed by politicians. A particular problem can also be analysed from many angles. As a result, a kind of synthetic re-configuration and re-contextualization of the available knowledge must take place, and many different points of view and disciplinary perspectives must be incorporated into interdisciplinary or transdisciplinary syntheses. The link to action and the possible consequences for policy recommendations frame the questions as well as the answers. In their narratives, experts have to act not only as if they know the answers, but also the conditions under which the answers will fit into an unknown future.

Narratives of Expertise

There are three important characteristics of narratives of expertise: they are transgressive, collective and socially distributed (and self-authorizing). These narratives are transgressive in two senses. First, they must address issues which can never be reduced to the simply scientific and the purely technical. The practices which they struggle to describe are characterized by overlaps and interlinkages. Unpredictable 'seamless webs' produced by the intermeshing of society and science (and technology), values and politics hold them together. To have any predictive value expertise must be able to understand the interlinkages that situate and bind the various relevant practices together. The second sense in which narratives of expertise are transgressive is that they address audiences which are never composed entirely of fellow-experts. These narratives have to be sensitive to a wide range of demands and desires and also relate to the heterogeneous experiences of their mixed audiences. The more audience-oriented expertise becomes (with the media acting as one of the main drivers of this audience-orientation), the more transgressive its narratives must be to reach out and speak to these diverse audiences.

Second, narratives of expertise are inherently collective. The competence of the individual expert is inevitably limited. So competence must be derived from a collective pool of expertise. The media, of course, prefer to personalize expertise by highlighting the 'star expert', and do not glamorize the work of committees. But, in practice, experts nearly all work in wider teams. Individual authority ultimately depends upon collective authority. Although in legal proceedings experts may be called on as individuals, they are testifying in the name of collectively validated expertise, rather than expressing only their own individual (or idiosyncratic) expertise. The narrative of expertise speaks in a collective voice, which has to be orchestrated with care in order to produce the maximum degree of consensus. Dissenting voices, provided they have a sufficiently good case, are permitted. But in order to continue to share in the collective authority, these dissenting voices must leave sufficient common ground. That is why consensus among experts is so important. Collectively expressed expertise is not addressed primarily to other experts (who might accept, and even value, robust dissent) but to lay audiences interested in concrete outcomes rather than abstract solutions. As a result, when experts (approximately) agree, they gain in authority.

Third, the scientific authority conveyed by expertise is now socially distributed. A good example of a body that has to engage with all kinds of heterogeneous expertise is the Human Genetics Advisory Commission in the United Kingdom. At first sight this is difficult to understand because social distribution appears to be inconsistent with the image of science speaking with one authoritative voice. To employ experts with high scientific reputations may still be considered useful.

But it is impossible to base the authority of expertise solely on the scientific reputations of these experts. Scientific and technical expertise is always open to contestation – and in circumstances outside the control of experts. Nevertheless expertise clearly is imbued with authority. What is the basis of this authority? Expertise has to bring together knowledge that is widely distributed, highly contextualized and heterogeneous. So the authority of expertise arises not from any specific site, nor from what single scientific disciplines or groups of successful researchers have to offer. Instead it rests on its ability to orchestrate many heterogeneous and context-specific knowledge dimensions that are involved. So the scientific authority of expertise is derived from the links that bind it together in its highly distributed form. Seen in this light, scientific and technical expertise is self-organizing or, better still, self-authorizing. For the narrative to be predictive, it depends on successful self-organization and self-authorization.

However it is these three characteristics of narratives of expertise – their transgressive, collective and socially distributed/self-authorizing qualities – which increase expertise's vulnerability. These narratives are public representations, which are observed, recorded and transmitted by the media. As such they are addressed to wider audiences, the reaction of which influences the effectiveness of their efforts to communicate. Transgressivity is open to contestation because, by definition, experts speak about matters that transcend their competence. This explains recurrent, but futile, attempts to re-draw boundaries in order to prevent their being trespassed. 'If only experts would stick to what they responsibly can claim to fall within their competence', 'if only they would speak as scientists and not venture into politics or economics' – such exhortations can still be frequently heard in the vain attempt to protect the authority of scientific expertise against the perils of transgressivity. But the confusions and complexities of the social world doom this attempt to failure. They confound and defeat every effort to re-draw the boundaries once and for all.

Similarly the collective nature of expertise can turn into a liability if the lonely voices of dissenting experts make themselves heard. The defeat of one opponent by another is in the nature of scientific controversies; as such, it is routine. But defeat of the majority by a few dissenters transforms that defeat into a great victory. However, the prevailing logic of the narrative of expertise suggests there are good reasons for erring – if erring cannot be avoided – on the side of the majority. Thus, the narrative of expertise is likely to remain open to the occasional victory by a dissenting minority. But where trust in scientific experts is already low, the public may become convinced that no experts are to be trusted, so accelerating the downward spiral of distrust. However, it is an equal mistake to rely too much on the authority of science, when only its outward symbols are retained and its substance is undermined. The narrative structure of expertise, therefore, is a forceful reminder that expertise is not only highly context-dependent, immediate and concrete in the structures of its production, but also in its power of persuasion. It is loaded with political and economic interests, riven by fundamental value-conflicts and inevitably short-term in its outlook. The distributed nature of the scientific authority of expertise makes for a precarious balance which has to be endlessly redressed and re-calibrated.

Future Expertise

Many thoughtful proposals have been made about how the present shortcomings of scientific-technical expertise can be remedied. But this will not be easy, because, as we have argued, current discontents with the authority and effectiveness of scientific-technical expertise are the result of two separate, although interrelated, processes. The first, linked to the rise of the individual-as-decision-maker in the dual role of consumer and citizen, has led to apparently irreversible changes in the decision-making structures of liberal Western democracies. The second, equally wide-ranging and also probably irreversible, has led to the emergence of a socially distributed system of knowledge production, which is no longer confined to institutional monopolies or knowledge bases organized along strictly disciplinary lines – in other words the emergence of Mode-2 knowledge production operating in a Mode-2 society. This process has further increased both expertise's context-dependence and its transgressivity, and stimulated the growth of a self-organizing although precarious system in which scientific authority itself is widely distributed.

The transformation that the narrative of expertise has undergone in the shift from modernity to the present age of late modernity implies a crossing of domains. The idea of expertise originated in the public sphere, where it was certified and regulated by the state, whether through the relative autonomy granted to the professions or a state bureaucracy. So scientific and technical expertise was an indispensable support for the expansion of the bureaucratic state. But, although expertise was allied to the growth of state power, its authority was not derived from the state. Instead it spoke in the name of Nature and represented the unequivocal and uncontested public voice of science. However, when expertise entered increasingly the private lives of individuals as consumers and citizens, the public voice of science had to contend with private whispers of citizen-consumers. These rival forms of localized experience and particularized expertise have extended beyond the private sphere into the modern *agora*. They now demand to be co-ordinated in order to attain an authority which transcends the aggregate of particular individuals. To some extent, this process involves the de-professionalization of expertise and a re-appropriation of expertise through lay participation. As a result science may now define itself as 'citizen science' (Irwin 1995), as 'advocacy' or even as 'participatory' science, in an effort to assert its openness and willingness to accommodate to the new situation, although the mechanisms of participation have still to be defined. In a Mode-2 society the existence of the *agora* indicates that the boundaries between public and private are overlapping, criss-crossing and shifting. It is in this sense that expertise is transgressive, collective and socially distributed.

This partial privatization of expertise, which is reflected in a decline of deference towards science and in demands that science should be more responsive to public expectations, can be regarded as a process of emancipation. But the diffusion of expertise is not unproblematic. A key issue is quality control. To be effective the outcomes of scientific and technical expertise must be correct, reliable and socially robust, regardless of their origins and processes. Its predictive power, rightly, must be assessed – at least retrospectively. But there is a further difficulty: the number of those who now judge the effectiveness of expertise and who assess whether decisions were 'correct' has been considerably enlarged. Any such judgements are open to interpretation and, consequently, to challenge. As with Mode-2 knowledge production, scientific quality alone is no longer the sole or sufficient criterion of expertise, even if it remains important. Additional criteria enter. Further complications arise because of the many possible ways in which knowledge can be linked to action. It increasingly depends on the context and the questions asked and who has been involved. Participation can only mean multiplication of decisions – with all their consequences. This intense and inherent heterogeneity makes quality control an urgent issue – but also an unresolved, and perhaps unresolvable, issue. The difficulty of transforming knowledge that is merely reliable into the socially robust and context-sensitive knowledge discussed in chapter 11 applies equally to the expertise which is based on such knowledge. However, one prediction is safe – there is no realistic prospect of any scientific body, however eminent, re-asserting a claim to monopolistic control of scientific authority without challenge. Scientific elites can no longer hope to restrict potentially open-ended demands for expertise or to restrain its inherent transgressivity. In the United States the experiment with 'science courts' has failed, essentially because scientific expertise cannot be treated in the same way as legal 'facts'.

What then can be done – or does anything go? One proposal is to install groups of pluri-disciplinary experts on a permanent basis (Roqueplo 1997: 41). Then the procedure of arriving at a reflexive interdisciplinarity would be as highly valued as the product of the deliberation of experts. This reflexivity would ensure that the bias inherent in scientific expertise, and its potential for producing, as well as resolving, conflicts, were acknowledged. Reflexivity always contains the hope that sensitivity to limitations is heightened. A permanent public debating space could then be established and a kind of reflexive confrontation institutionalized as a central and positive element within it. The creation of such an expert system might encourage longer-term perspectives, which make expertise less vulnerable to the pressure of the 'immediate'. But such a system might not go far enough to address all our discontents.

Just as a socially distributed Mode-2 knowledge production system has profound implications for how scientific and technical expertise are conceptualized, so a future system of expertise must recognize these implications by becoming more open and inclusive. The basis of knowledge and expertise must somehow transcend the narrow institutional confines of traditional scientific institutions. To remain effective in the twenty-first century, expertise must rely on both its public and private forms, as they emerge and diffuse in the *agora*. Experiments are needed to explore how these two forms of expertise might be reconciled. Lay participants and their experience must receive a recognized place – but lay participation is not the whole answer. A socially distributed and transgressive expert system needs to create and nourish a truly pluri-disciplinary knowledge base, which in turn can develop transdisciplinary methods of translating knowledge into action. This is where the dimension of time enters the argument. Paradoxically, immediacy of response can only be guaranteed if an expert system adopts a long-term time perspective. Only if a longer-term perspective is adopted will the system be able to draw upon more knowledge and expertise than is currently available in order to choose that which is most immediately appropriate.

The narrative of expertise in a Mode-2 society needs to acknowledge its constituent features – transgressivity, collectivity and the socially distributed and self-authorizing nature of its authority. The basis of legitimacy can no longer be derived from outside. Instead it emanates from the actual process of knowledge production. So trust cannot be taken for granted or regarded as given; it must be continually re-established. Heterogeneity has now become a pervasive characteristic of knowledge production, whether in the guise of science (or research) or expertise. Perspectives, questions and answers, interlinkages between private and public, are all heterogeneous; so too are the social actors involved. As a result expertise is successful at identifying dilemmas – but, because of this success, cannot reduce their inherent ambivalence. Perhaps the insights of anthropology may help here. In the words of Marilyn Strathern, 'anthropological practice always was the expertise of being in two places at once, here and there, seeing "ourselves" and "others" at the same time. It depended on the perception of a divide that was constantly dissolving and reforming' (Strathern 1996).

To be successful, expertise must be able to predict (because success implies prediction). But in late modernity expertise can only be successful if these features – including their unresolvable ambivalence – are recognized as inherent to it. These characteristics must also be made explicit. In the early phases of modernity, scientific and technical expertise was embedded in a co-evolving process of modernization. The state was the central actor; science and technology were regarded as forces which impacted on society but were 'external' to it. In late modernity all this has changed. The state is no longer such a central actor and the vacuum created by its retreat has been filled by the chaotic forces of globalization. Partly as a result expertise has become disembedded from the public realm and distributed through society. Its 'privatization' can be regarded as a re-appropriation of expertise through lay participants because even lay people can be sources of expertise. But this leaves open the question of how a new, publicly acceptable expertise can grow out of the process of social diffusion and how its legitimacy can be established. Is it possible for scientific and technical expertise to be re-embedded in wider social contexts when the interlinkages between these contexts are inherently volatile? Is such expertise, although no longer in control of the future, nevertheless able to offer a kind of time-released wisdom which can anticipate future uses and consequences without being able to predict them?

15

Re-Visioning Science

In this penultimate chapter it is necessary to revisit the co-evolutionary trends that mark both knowledge production and socio-economic change in a Mode-2 society – in particular, the production of uncertainty. In this context the 're-thinking' of science can also lead to a 're-visioning' of science, to another image that is more consistent with actual scientific practice but need not be devoid of normative content. The traditional claims of science to a special epistemological status can no longer be maintained. Science can no longer base its cognitive and social authority on a claim to have unique access to the order of the natural world (the understanding of which takes precedence over understanding the social world, inherently more messy because it is contaminated by social interests, value judgements and ideologies). Assertions about the autonomy of science will not carry much weight in the *agora*, nor will the insistence on its inherent universalism and incontestable objectivity. The validity of such claims partly depends upon science being constituted through a distinctive *praxis*, which is different from other social practices. We have attempted to show how science, because of the increasing contextualization of knowledge, has become embroiled in the *agora*. Instead of being clearly demarcated from other forms of social practice, and far from being uniform or unified, science itself now consists of a set of complex practices, deeply embroiled, integrated and implicated with society.

At times these scientific practices share common characteristics with other social practices – such as common patterns of communication, similar principles of work organization or shared competitive behaviour. Viewed in this way there is nothing special about science; scientists simply observe, calculate, theorize and model like everyone else, even if they are cleverer and more careful. But at other times, scientific practices, although still infused by the social, are far removed from everyday life and common-sense understanding; they seem to be bizarre and strangely cut off from the common world. But as Simon Schaffer has argued, both accounts – the everyday and the arcane – are schematic. They represent recurrent oscillations explicitly designed to account for science's remarkable efficacy. They are visions as much as images, reinforced and rendered concrete in exhibitions and public displays. They provide the important settings in which such images are forged, and the claims of scientists for the efficacy of the sciences sustained, under changing circumstances (Schaffer 1997).

Today, science still appears *par excellence* as a problem-solver or trouble-shooter. For example, in public debates about the degradation of the natural environment it is constantly argued that 'solutions' can only be produced by an even stronger commitment to, and by, science in order to monitor, measure and model the complex interactions between natural processes and socio-economic ones. Only increased research effort (and the required resources) can produce 'solutions' – a notion, admittedly, that in the environmental field has had to be stretched to include the considerable political, as well as scientific-technical, inputs into sophisticated policy-making and implementation at every level from the local to the international. Or, to take another example, in the hotly disputed public controversy about genetic engineering and biotechnology's expanding powers of manipulation, science's defence emphasizes its potential, and potent, contribution to finding cures for incurable diseases or 'solving' the world's famine problems.

Science's appeal for greater public support (including more public funding) on the grounds that it works and can overcome the problems of the future is not merely cynical – despite the occasional and counter-productive hyperbole. Nor is it merely a cynical posture or a tactical response to the public's own perceived cynicism. The grip of the 'Utopia' of the sciences on the collective imagination, one of the few that has survived to the end of this 'short century of extremes', in Eric Hobsbawm's phrase, is still strong, even if it is now expressed in less native and technocratic guises. The belief is still widespread that there is no problem which at some point is not amenable to a scientific solution or a technological fix – even if such solutions are nowhere on the near horizon and even if the term 'solution' has to be interpreted in more modest and localized terms. Nor is this belief without warrant.

Visions, images and beliefs cannot sharply be demarcated from knowledge. Knowing is inevitably rooted in some set of beliefs. So science, among many other things, is also a belief system. Far from being dangerous illusions or Utopian projections, visions are a precious resource, an intangible asset, that may help to provide the necessary (but typically missing) link between knowing and doing. But to appreciate their significance it is necessary to understand where these visions of science originate, how they are shaped and by whom. It is important to recognize how visions and images interact and also how wide the gap separating images from practice can become before an uncontrollable backlash is provoked. It is no longer possible to adopt a naive 'modernist' account of the efficacy of the sciences, of their origins, generation and outcomes. Such an account is also a narrative in which agency, mistakenly imputed to the main actors of the narrative, is actually constructed through the narrative. Instead of being prime movers, scientists are often portrayed, or see themselves, as the main actors, emerging in a more reflexive version as characters in a multi-authored story.

In our (different) account of the altered practices of knowledge production as it has moved beyond the heterogeneous sites which represent the context of application into the even wider arenas of implication, we have argued that these changes must be understood in terms that are both more reflexive and more realistic. Of course, scientific autonomy must be preserved. It is a precondition for the formation and perseverance of scientific identities and therefore an essential precondition of scientific creativity. However, autonomy plays a similarly positive role in stimulating many other kinds of human creativity. Scientific autonomy, therefore, will take on highly localized forms; it will have to be justified in each case and for each individual research project. In the same way scientific objectivity must be preserved, because it is also an essential precondition for the production of reliable scientific knowledge. But again, it will be transformed beyond recognition. There can no longer be universal scientific objectivity – or only at a highly abstract, and practically meaningless, level. There can no longer be established canons of rules which must be followed in order to guarantee scientific reliability. Instead, scientific objectivity will have to be re-defined to become localized and contextualized; it will have to be shaped to anticipate the specific contexts where it will be challenged. But objectivity will be successfully re-asserted, if knowledge produced is socially robust. To be socially robust, though, this knowledge will also have to anticipate the many heterogeneous contexts of implication which are an inherent feature of knowledge produced in different contexts of application. Instead of relying on foresight exercises that scan the potential contexts of application, socially robust knowledge builds on the epistemological foresight of its context of implication.

The emergence of a Mode-2 society in co-evolution with Mode-2 knowledge production demands a historically unprecedented openness on the part of science. Merely to add to a supposedly 'hard' scientific core additional outer layers consisting of 'softer' institutions, 'softer' norms and 'softer' behaviour on the part of scientists, all of which are designed to give greater weight to economic and social issues such as value-added wealth creation, or ethical concerns, cannot work. Instead science should attempt to reconstruct its authority and image according to the fluid and transgressive dynamics of a Mode-2 society (in which it is already operating). This reconstruction must take place on many different levels. If science is in reality much more heterogeneous, diverse, local and disunited than its still-dominant image suggests, the mechanisms through which, despite this apparent disunity, strength and coherence nevertheless continue to be generated must be understood (Galison and Stump 1996). How can scientific identities retain disciplinary affiliations while, at the same time, embracing new transdisciplinary modes of working? And, if politicians and policy-makers now emphasize 'research' rather than 'science', how much 'science' is still needed to produce particular kinds of 'research'? If the much vaunted hard epistemological core of science turns out to be empty, let it be empty. What has replaced this core is more interesting – shifting local practices, consisting of heterogeneous clusters of beliefs, inextricably interwined with the materiality of science, with methods and procedures, local practices which place different values on objectivity, proof and verification and combine them in different ways. These local practices are more or less robust and reliable depending on context; they comprise a medley of theoretical, instrumental and experimental elements; and they are wrapped up in a social epistemology that equips science to move at ease in the *agora*.

But how can science ever share its cognitive and social authority? In the Platonic/Socratic *agora* the people were held to be incompetent and unqualified. Viewed from an elitist-aristocratic perspective the people could never become competent and qualified, so they had to be excluded from government. In the prolonged struggle for democracy in Europe, the advantages, now obvious, of having the representatives of the people argue with words in parliaments, instead of killing each other or plotting to overthrow the government, were only gradually realized. The same has happened to science. Even as late as the 1930s, when a small circle of young, mainly Marxist, scientists in Cambridge debated the future of science and society, their vision was still essentially elitist. Despite their political sympathies they still believed that science would lead the way, that science was predestined to guide society (Haldane 1923/1995). At much the same time members of the Vienna Circle were engaged in a similar debate, although it took a different turn. They believed that in order to bestow democratic competence on the people, the scientific method had to be purified; all personal elements and differences, which they saw as leading to obscurity and, eventually, to a dangerous irrationalism, had to be eliminated. Anyone would be able to do good science – and in the true spirit of the Enlightenment, everyone was eligible – provided they followed these precepts and a method so impersonal that even language, as the ultimate source of contamination, had to be curbed.

In practice, as the members of the Vienna Circle knew because many lectured in evening classes for workers, the situation was very different. Access to education was not equally distributed. Only Ludwik Fleck, writing in Poland on the margin of what was then considered to be the scientific world, argued strongly for the overlapping of different 'esoteric' and 'exoteric' communities. He took the discovery of the causes of syphilis as an example, retracing the 'genesis of a scientific fact' and demonstrating that it had been deeply immersed in contextual factors. He devoted much attention to the 'circulation of ideas' and how they were transformed by being shared between different kinds of communities, which he aptly called *Denkkollektiv*, including lay people as well as scientists (Fleck 1935/1979). Today, as a result of the further inroads that the process of democratization has made into Western industrialized societies, mutual adjustments and appropriations have taken place between science and society. But science still sees itself as a meritocratic, not a democratic, institution. In common with other major institutions in society, science has been confronted by a decline in its authority, both cognitive and social, and in public trust. Participation in education has reached new heights. Science now finds itself interrogated by the spirit of criticism – the extension of which is an outcome of educational improvement but which is also so central and dear to the scientific system.

Of course, differences between experts and non-experts remain. So do levels of competence, skills, originality and the propensity to take risks, so crucial for creativity. But, as we have argued in the last chapter, expertise itself is becoming distributed and transgressive. There is nothing to suggest that science – or any other institution in society – will be able to recover its lost cognitive and social authority.

Too many of the parameters that built and sustained such authority have changed. However, this does not mean that chaos and disorder are inevitable. The 'flight of science from reason' (to borrow the title from the proceedings of a recent American Academy of Sciences (AAS) conference) is not taking place – despite apocalyptic warnings to the contrary. Nor is scientific quality likely to collapse. On the contrary, more verification checks are now made and higher standards of scientific behaviour are demanded. More and more ethics committees are being established and anti-fraud measures devised. Auditing, the most recent expression of self-observation, self-monitoring and self-organized accountability, is now pervasive. Nor is there any evidence that science is about to sell out to the profit imperative of the global economy;, instead the state is having to re-evaluate its regulatory duties as well as its obligation to provide public goods that cannot easily be substituted – in research and science policies as elsewhere.

Scientists can reconstruct their authority – and the image of science – in a Mode-2 society by understanding (and communicating) better how that authority is constituted, how images are made and how trust is maintained. First, they must recognize that they cannot rely solely on what they know as scientists because social knowledge is also involved. In the process they may discover how much of this social knowledge is already present in what they do. Instead of leaving it to social scientists to analyse what scientists do or arrogantly assuming that scientists know everything already, they need to engage in a creative dialogue with social science. Second, scientists must overcome their fear of contamination by the social. Any attempt to rethink the place of people in knowledge must include scientists; they too are part of the social. The subordination of the subjective in search of objectivity from which all subjective traces have been purged, and also perhaps the repression of wilder elements of the scientific imagination, have led to a reluctance to recognize the Other in ourselves. Third, by recognizing how much scientific work has already been transformed by the process of contextualization, scientists may discover a New Frontier. However, this frontier will not be reached through foresight of possible applications, flowing in linear fashion from basic research. Rather it lies in the messy and convoluted interstices that link scientific activities in multiple ways with the many local and social contexts with which they interact. The new frontier leads back to where science began – to the social processes that first led modern science to discover Nature and to learn how to manipulate and master it. But science also had to acknowledge the role that people and the social world played in this process of discovery and mastery, which opens up a frontier full of potential but also of new constraints.

Institutional Aspects of 'Re-Thinking' Science

The identification of recurrent co-evolutionary patterns in knowledge production and in a Mode-2 society suggests the existence of many implicit mechanisms that lead towards greater 'integration' rather than maintaining a model of 'segregation'. The inherent generation – and growth – of uncertainty is an example of such a co-evolutionary pattern. It develops out of an increasing awareness that the number of questions greatly exceeds the number of answers that can be given in a finite time. Because answers give rise to further questions, the horizon of possible investigations expands more or less continuously. The range of possible directions for research is unimaginably large, which raises the question of how we can choose which to follow. This has practical as well as theoretical relevance. For example, in research careers, choices have always depended on pursuing the 'right' questions at a given time and under given circumstances. However, there are now many more uncertainties and so risks. As a result some researchers hedge their bets by embarking on several different lines of research at the same time; others may choose to deploy their skills in other environments and develop second strands of interest and expertise, for instance in the expanding world of consultancy.

Consequently the uncertainty generated by the plethora of choices and decisions appears to promote a greater variety of experimental behaviour. This, in turn, produces institutional adjustments which may be perceived by others as a serious 'disturbance of the local balance' and so provoke countervailing actions, which generate further insecurity and uncertainty. On the other hand, in such situations local imbalances can acquire a degree of self-organization; enclaves are established that provide support to their members in an increasingly self-reinforcing manner. However, there are limits to such piecemeal increases in institutional permeability; the accumulation of institutional adjustments forces the institution as a whole to become more flexible. This pattern can be observed in the development of commercialization activities in many universities. At first commercialization was a response to declining public funding, but later institutional developments have both shaped and been shaped by academics responding more radically to an uncertain future.

The theoretical consequences of similar adjustments have been analysed by Luhmann and have led him to contend that 'modern society is shocked by its risks' as it struggles to cope with an overload of decision-making. Within the framework of systems theory, Luhmann describes how decisions have to be 'made' because of the way in which modern society sees itself. Decision-making is needed when past states and future states no longer 'match' automatically. 'Matching' can be seen as either continuous or discontinuous. In the first case things stay as they are. In the second the process of change is interpreted in accordance with nature ('getting old' or 'decaying'). Decisions could be interpreted in the same way. They produced discontinuity, but in accordance with natural ends – whether 'erroneous' or 'sinful'. However, modern society abandoned this interpretative frame. It opted for discontinuity and the systematic and relentless pursuit of 'newness' – with modern science as its *avant garde*. As a consequence, the temporal structures of society changed. Planning horizons replaced the rhythms of the weekly markets held in different villages. Technological developments need 'their' time and it is difficult to foresee whether they will succeed 'in time'. But, above all, the merciless temporal regime of the machine that came to dominate the industrial revolution left its indelible stamp on working life and life as work (Luhmann 1996).

Only recently has society escaped this iron regularity – but only to be confronted with what had not been foreseen. Instead of technology 'liberating' people, it has merely enabled them to do more things. Far from facilitating decision-making, it has opened up a range of new choices. This has led to a state of near-exhaustion in the lives of many people, which has been described variously as 'hurry sickness', 'time famine', the search for the 'unique moment' or 'time deepening' which seeks to intensify the experience of time or simply, a better 'balance' between a hurried life and work. The explosion of new information and communication technologies has accelerated and intensified these trends. Decisions also have their own temporal frames. They are not simply 'good' or 'bad' choices, according to subjective or objective criteria. Decisions try to find alternatives to the present and to create a structure for the future. They cannot determine a future that is open and uncertain, but they can (and do) project a goal or an objective onto this open horizon. For Luhmann these projections are much more than preferences; they are the 'difference' between a preferred state and what would be the case without interventions. Decisions thus introduce newness – and with it comes 'the unavoidable unity of fate and risk' (Luhmann 1996: 13).

New forms of economic rationality, which have already been discussed, have also shifted the focus from the material conditions of production to a more abstract and speculative approach towards both economic exchange and investment. For example, according to the new field of the quantitative assessment of financial risks, current research investments provide information which enables better use to be made of future investments. Viewed like this a research investment becomes analogous to a financial derivative. A research investment is treated as the cost of keeping an option open until the decision to go ahead with or cancel the project can be made with greater certainty about the outcome, which is produced by the findings of the research project. As a result, conventional logic about short-term investment and profit calculations is displaced onto this higher level of speculation – with both productive and disruptive results. Already both the 'old' and 'new' economic rationality have had a considerable impact on the conduct of research and the transformation of the traditional 'collegiate' culture inside universities by the advance of the 'audit culture' and the diffusion of managerialism.

Intellectual property rights contribute another secondary 'economic' effect and another forward-driving rationality to ideas, artefacts and technologies which are already being shaped by creative social processes. Their anticipated utility, their inherent potential for further utilization and transformation, are explicitly recognized in legislation and contracts which regulate access to benefits of future derivatives of their potential. Intellectual property rights do not arise from the process of using knowledge to gain more knowledge, which is considered to be an integral part of all knowledge production. Instead they relate to the specific know-how generated by making knowledge tradable, by turning it into something else which is economically exploitable and capable of generating future benefits. It is the potential that is privileged. The potential is derived from the original idea, invention or discovery in a series of steps each of which confirms its underlying materiality but also strengthens the dynamic of this economic rationality. In this way the original discovery or invention is granted an afterlife, value (and valuation) of which may be considerably greater than its direct primary benefit.

Another example of how economic rationality pervades scientific practice is provided by the largest collective enterprise in the biological sciences, deciphering the human genome under the auspices of the US National Human Genome Research Institute (and similar ventures in other countries). This enterprise has been accelerated not only by using automated sequencing and computer programmes (such as PHRED and PHRAP), but also by treating it as a multi-dimensional optimization problem. The most remarkable result has been that research problems are now tackled at the same time as production problems, rather than following an old logic that data production must precede genomic research. The enhanced importance of expectations and potential, of which this is a good example, is one of the co-evolutionary elements that shape both the development of knowledge production and the emergence of Mode-2 society. Expectations invade the process; potential drives it forward; and anticipations reduce the delay between plans and actions. Although appearing like an assembly line, the sequencing pilot centres which comprise the collaborative network have been allowed to choose their own strategies. The sole aim has been to achieve higher production rates. In the process the culture of biological research is being radically transformed. As a result the demand for sequenced genomes will continue to grow, even after the expected completion of the first human genome sequencing in 2003, two years ahead of previous projections.

The self-organizing capacities of both Mode-2 knowledge production and Mode-2 society are also crucially important. Universities, in particular, have responded to the general shift towards an internalization of social control and adapted to an 'audit culture' of accountability. Audit topics and audit methods are intertwined. Michael Power has observed: 'What is audited is whether there is a system which embodies standards and the standards of performance themselves are shaped by the need to be auditable... audit becomes a formal "loop" by which the system observes itself' (Power 1997: 36–7). In a similar way research which is particularly exposed to public anxiety has espoused safety guidelines and acquiesced in a host of regulations and even restrictions in its eagerness to anticipate and accommodate demands for greater public accountability. From the very start the examination of the ethical, legal and social implications of genome research was an integral component of the US Human Genome Project. The Ethical Legal Social Implications (ELSI) programme has generated a substantial body of scholarship in the areas of privacy and fair use of genetic information, safe and effective integration of genetic technologies and information into healthcare, and other educational and policy issues. This work is expected to continue. In such ways a reflexive element, an increased awareness, not only influences individuals and informs their behaviour; it also operates on an institutional level. However, there remains a danger that any call for greater 'reflexivity' is reduced to a mere 'after-thought'. For example, proposals to establish ethics committees may be agreed without serious discussion because this seems to offer an easy 'ethical fix', a reprieve from further public scrutiny and a shifting of responsibility to another agency.

Normative Aspects of Re-Thinking Science

This brings the argument back to the normative content of this attempt to 're-think' science. There are many good arguments, as well as solid empirical observations, to support this attempt. The contextualization of scientific knowledge production has to be taken seriously; the context of implication must be addressed as well as the context of application; scientific knowledge must be not only reliable but also socially more robust. However, our account of the co-evolution of Mode-2 knowledge production and Mode-2 society also contains normative elements. Amartya Sen has argued, passionately but reasonably, that democracy is not incompatible with economic growth. In a similar vein, science and its core values have become compatible with the demands of the *agora*. According to Sen, three functions are ascribed to democracy: first, political freedom; second, the instrumental support it grants people in their efforts to express their claims to political attention; and, third, the opportunity to learn from each other. Even the idea of 'needs' (including an understanding of 'economic needs') requires public discussion and exchange of views. In this sense democracy has a constructive importance, in addition to its intrinsic and instrumental value (Sen 1999). In this third respect – the constructive importance of democracy in making people more and better aware of their 'needs' – science can also make a constructive contribution, provided that it is ready to enter the *agora* and is willing to listen as well as to engage in an authentic multi-logue. Many scientists, however, cling to the tradition of the Enlightenment and, when a gap opens up between science and society, still pin their hopes on improving the 'public understanding of science'; lay people must see the world through scientists' eyes or, if they cannot, trust scientists. But it is well known that more information and 'understanding' (especially in its more abstract, context-independent forms) does not generate greater empathy or increased comprehension. On the contrary, better education increases the readiness as well as the ability to ask critical questions – for example, about traditional divisions between 'experts' and 'lay people'.

Science has much to learn from and much to contribute to an enlargement of democracy when it is ready to move without fear in the *agora*. Its success has demonstrated its ability to embed itself into diverse contexts. Knowledge production has already become highly contextualized. But certain rules must be observed. The need to establish norms of behaviour arises when 'internal' actions have 'external effects', including the production of public goods or public bads. Norms are also needed in cases where markets cannot easily be established or transaction costs are high (Coleman 1987: 140). The Mertonian norms, which claimed to regulate the behaviour of scientists and to sustain an ethos of science that was functional as well as indispensable, are now largely an anachronism. Part of the reason is that they have become obsolete; part is that the gap between norms and actual behaviour is now too great. Even in their breach, rules must maintain a safety margin.

But this does not mean that science has been plunged into a normative void. Current norms are improvised, diverse, volatile and even at times contradictory. This is not necessarily alarming. But the public image of science has not kept up with the many far-reaching changes which have transformed the work and life of the majority of scientists. Already the expectations and prospects of younger scientists are radically different from established patterns of life-long tenure enjoyed by older generations. The increasing presence of women in science (although less in engineering) is beginning to modify a predominantly male culture. But whatever the indelible and irreversible marks these changes inside the science system have left on the future outlooks of scientists, on their norms and values and, above all, their research practices and life-styles, they have still not been communicated to a wider public. So the public image of science has lagged dangerously behind its contemporary practice.

Public images of science are not shaped by any particular agency. No one is responsible. Although the media play a decisive role, they follow wider trends in shaping the images they project. The school system, despite some pioneering efforts to the contrary, is still dominated by an out-of-date 'two culture' model, which regards science as essentially unproblematic (at any rate, in its 'internal' constitution) and the social world as overwhelmed by its problematics. Museums and other public exhibitions, which should be ideal sites to mediate between science and arts, do not always take full advantage of these opportunities. Universities, in the throes of radical transformation, are too self-absorbed to become the shapers of an alternative (and more realistic) image of science. Research councils are obliged to develop strategies to implement complex, and sometimes contradictory, research agendas. Governments pursue foresight exercises in order to sharpen their priorities for social policy and wealth creation (while ignoring wealth distribution). The Fifth Framework Programme of the European Union is perhaps the most ambitious of all by seeking to reach out to Europe's citizens as 'users'. But the programme is thoroughly bureaucratic and the commitment of scientists is half-hearted.

So where can a re-visioning of science take place; how can a different public image be constructed? We have repeatedly emphasized that many changes occur in the processes, and patterns, of collaboration and that new, sometimes quite unexpected links arise in the interstices between disciplines and research fields or across institutional and other boundaries. Individual researchers and research groups – and even people outside the research system proper – are now busy setting up collaborative ventures. They might never have met but for the profound changes in knowledge production which we have described. Forms of Mode-2 collaboration range from the setting up of multi-disciplinary centres, which are designed to bring the physical and biological sciences closer together, to new curricula which will mix biology, physics, chemistry and mathematics in novel ways. They include new organizational forms, such as the Co-operative Research Network of sequence production centres. The network will be funded with $70 million a year until 2005 and has been given clear targets about how many mega-databases are to be produced – but at the same time has been encouraged to foster interaction among the networked laboratories. Other examples of Mode-2 collaboration include the many and diverse forms of interaction between university research groups and industry – including the help and guidance given to graduates who want to try setting up their own firms. It includes many fascinating attempts on the micro level, dispersed around the world, to bridge art and science in novel and imaginative ways. In short, Mode-2 knowledge production teams and collaborative organizational forms have not only multiplied but endlessly spawn new variants.

Of course, it may be argued that the effects of the contextualization of knowledge have been exaggerated by concentrating on cutting-edge or *avant garde* research (and the changes in the values and careers of younger students and graduates). There is substance in this counter-argument. Just as Mode-1 science is still alive and well (especially in universities), so knowledge production in a Mode-2 society is still in the process of expanding and spilling over from the innovative niches where it began. But there is a historical precedent. During the Renaissance the intellectual elites of Europe decided that the greatest thoughts had been those of a handful of thinkers in the city-states of Greece and in ancient Rome. They concluded that patient reading and understanding of these ancient works was essential for understanding and building their present. This was made easier by remarkable interlinkages between different social groups.

The result was a new literature, a new art, a new architecture and new technologies, which blended together with elegant ease. Scholars and artists began to collaborate, exchanging their reading and interpretation of practical mathematics and ancient surveying methods, of Latin orations and poetry. Artists ceased to see themselves as mere craftsmen and became aspiring scholars in their own right. Their political and material environment was transformed to provide the economic, as well as political, space for the kind of patronage under which the new culture flourished. As Panofsky had argued some time ago, the culture of the Renaissance was made possible because the barriers between what had been sharply separate cultural realms, like the visual arts and the study of natural philosophy, suddenly fell. As a result artists could claim a new intellectual status, since their mastery of anatomy and perspective gave them an understanding of nature that surpassed that of scholars who had claimed to be their social and intellectual superiors (Grafton 1999).

There was a similar transgression of boundaries in the scientific revolution of the seventeenth century and the industrial revolution of the nineteenth. In the first case two separate worlds, speculative 'natural philosophy' on the one hand and on the other, proto-scientific artisanal handicraft and expertise, with its emphasis on experimental techniques and instrumental technologies (which were often difficult to tell apart), were powerfully combined. In the second case traditional demarcations between science, capital, technology and industry were transgressed. Pioneers of the industrial revolution were all four – scientists, capitalists, technologists and entrepreneurs. Modernity itself was produced by the transgression of the traditional categories of pre-industrial, pre-urban and pre-secular society, and the inventions of new and – by contemporary standards – dynamic and volatile categorizations. The co-evolutionary trends which we have identified in both science and society – the increasing contextualization of knowledge production and the emergence of the *agora* – may represent another such transgressive moment in human history – and provide an equally creative opportunity to re-think science (and much else besides).

However, our modest attempts to re-think science do not necessarily herald a renaissance of science. On the contrary, the current fashion for 'new visions' for science betrays the persistence of one-dimensional and technocratic views. These so-called visions are often dominated by attempts to 'naturalize' mind, matter and life. Maybe reductionism writ large, not as actual practice but as ideological programme, is determined to fight its last battle (Wilson 1998). These so-called 'visions' dangerously underestimate and neglect the creative potential of re-connections, of tearing down boundaries and sponsoring interaction between different social groups, of freely mixing scientific cultures with other forms and expressions of contemporary culture to ask once again the question that Renaissance Europe asked half a millennium ago: what has science to offer to life?

16

Re-Thinking Science is not Science Re-Thought

A Framework for Re-Thinking Science

The argument of this book is organized around the description of four interrelated processes. First, it is contended that the emergence of more open systems of knowledge production – Mode-2 science – and the growth of complexity and uncertainty in society – Mode-2 society – are phenomena linked in a co-evolutionary process. The implication is that not only does science speak to society (it always has), but that conditions are established in which society can 'speak back' to science. We have argued that the links between them can be understood in two different ways: in terms of the erosion of modernity's stable categorizations – states, markets and cultures – and also in terms of the transgressive, distributing effect of co-evolutionary processes, in particular the generation of uncertainties, the pervasiveness of a new economic rationality, the transformation of time and space and the self-organizing capacities of science and society. Both interpretations tend to the same conclusion: science and society have both become transgressive; that is, each has invaded the other's domain, and the lines demarcating the one from the other have all but disappeared.

Second, the process of 'reverse communication' is transforming science, and this, in the simplest terms, is what is meant by contextualization. Here, the contention is that contextualization involves not just an increase in the number of participants, their institutional or disciplinary affiliations, their experience, interests and networks, or even an expansion of the lines of communication between them. It also evolves in ways which we characterize as 'weak' or 'strong' and in which the shared definition of problems, the setting of research priorities and even to some extent the emergence of new criteria of what it means to do good science may be affected. But perhaps the strongest sign of contextualization occurring is the place accorded to people in the production of knowledge, through conceptualizing how people enter as users, as target groups in markets or addressees of policies, even as 'causes' for further problems to be tackled, – or as 'real' people in innumerable interactions and communicative processes, ranging from new modes of investment and financing for research, to legal regulations and constraints that shape the research process, to markets and media, households and internet users, other scientists and millions of sophisticated and highly educated lay people.

Third, the process of contextualization moves science beyond merely reliable knowledge to the production of socially more robust knowledge. To specify this process three main observations are advanced. The first is that the basic conditions and processes which have been underpinning the production of reliable knowledge are not necessarily compromised by the shift to socially robust knowledge. Reliable knowledge remains the indispensable 'condition *sine qua non*' of the fact that 'science works.' But if reliable knowledge has been undermined, it is possible that this has occurred as much by the narrow reductionism of scientific practice as by any attempt to widen the range of stake-holders or more systematically to articulate the context in which science is produced. Science was recognized to be inherently incomplete because it is primarily a method, but also because to achieve reasonable reliability the problem terrain had to be circumscribed and, therefore, restricted to the judgements not of the scientific community as a whole but of a peer group. Both of these aspects are carried forward into socially robust knowledge: knowledge remains incomplete, no longer only in the conventional sense that it will eventually be superseded by superior knowledge, but also in the sense that it is sharply contested and no longer entirely within the control of scientific peers. This shift involves re-negotiating and re-interpreting boundaries that have been dramatically extended, so that science cannot be validated as reliable by conventional discipline-bound norms; while becoming robust, it must be sensitive to a much wider range of social implications. The third observation is that the epistemological core of science is actually crowded with a variety of norms and practices which cannot readily be reduced to a single generic methodology or, more broadly, to privileged cultures of scientific inquiry. One outcome of these is that the sites of problem formulation and negotiation of solutions move from their previous institutional domains in government, industry and universities into the *agora*. The *agora* is the public space in which 'science meets the public', and in which the public 'speaks back' to science. It is the domain (in fact, many domains) in which contextualization occurs and in which socially robust knowledge is continually subjected to testing while in the process it is becoming more robust. Neither state nor market, neither exclusively private nor exclusively public, the *agora* is the space in which societal and scientific problems are framed and defined, and where what will be accepted as a 'solution' is being negotiated.

Fourth, the range of perspectives found in the *agora*, together with the ability of their proponents to articulate their wishes and concerns as well as to mobilize resources for research activities, implies a more complex role for scientific and technical expertise in the production of socially robust knowledge. The novel factor is that the role of scientific and technical expertise is changing as expertise becomes socially distributed. The immediate consequence of social distribution is the fragmentation of established linkages between expertise and established institutional structures whether of government, industry or the professions. The questions asked of experts, however, are neither the same as, nor simple extensions of, the ones which arise in their specialist fields of study. From the beginning experts have to extend imaginatively their knowledge to widely disparate areas, and try to integrate what they 'know' now with what others want to 'do' or have to 'decide' for the future. Narratives of expertise are constructed to deal with the complexity and uncertainty generated by this fragmentation. They have three characteristics. They are transgressive, told in a collective voice and self-authorizing. They are transgressive in the sense that experts must respond to issues and questions which are never only scientific and technical ones, and in the sense that they must address audiences which never consist only of other experts. They are collective in the sense that the limits of competence of the individual expert call for the involvement of a wide base of expertise which has to be carefully orchestrated if it is to speak in unison. Finally, the social distribution of expertise mirrors the social distribution of scientific authority. Since expertise has to bring together knowledge which is itself distributed, contextualized and heterogeneous, the authority of expertise cannot arise at one specific site, or from the views of one scientific discipline or group of highly respected researchers, but precisely from bringing together the many different and heterogeneous practice-related knowledge dimensions that are involved. The specific 'scientific' authority of expertise resides in the links that bind it together in its highly distributed form. In this sense, scientific and technical expertise is self-authorizing under conditions of Mode-2 society.

These four processes – the co-evolution of science and society in a Mode-2 direction, contextualization, the production of socially robust knowledge and the construction of narratives of expertise – are interrelated and will be brought together, often in conflictual and controversial forms, in the *agora*. But they also can form the main elements of a framework for re-thinking science. Co-evolution denotes an open, and certainly more integrated, system of science–society interaction which enhances the generation of variety, whether in the choice of scientific problems, colleagues or institutional designs, on the one hand, or the selective retention of certain choices, modes or solutions on the other hand. Success is often a determining criterion, but its definition is not given *a priori*; rather, it is made up in the process of selection itself. This is so even as these innovation-generating approaches, while responding to uncertainty and complexity, promote permeability, leading to more complexity and uncertainty, which encourage further innovations, both in the scientific-technical and social sense. Increasing permeability provides the basis for greater contextualization, by opening up the number of routes along which society can speak back to science. These processes, in turn, lead to the social distribution of knowledge, knowledge that is valid not only inside but outside the walls of the laboratory. As the walls of laboratories have become open, more and more scientists take their places as actors in the *agora* and so broaden the range of experts whose view might be sought on a particular problem or issue. The media partly play the role of indispensable allies, partly the role of critical opponents in this process. To cope with this a further development in the utilization of scientific and technical experts is needed. To overcome the fragmentation that arises from the social distribution of knowledge, experts need to realize the inherent transgressive nature of expertise, meaning that expertise always addresses more questions than it can answer. They also need to speak with a collective, although negotiated, voice which becomes self-authorizing, because there is no external authority available other than that guaranteed by the integrity of the process in which advice is tendered. Reliable knowledge can become socially robust knowledge only if society perceives the process of knowledge production to be transparent and participative. This, in turn, depends upon a reciprocity in which the public understands how science works but, equally, science understands how the public works. This enhanced mutual understanding needs to be guided by a vision, supported by appropriate images as well as transparency about how they have been generated and by whom.

The vision developed here is processual. It is an invitation to re-think science while emphasizing that even the best of 're-thinking' is not yet accompanied by the changes that depend on 'science re-thought'. It is the precise opposite of a blueprint which sets out the new 'arrangements' between science and society which are, to some extent at least, conceptually closed, awaiting only implementation. That the authority of science in the future will have to be established in an ongoing process that needs to be worked out again and again in each concrete situation is the meaning of the somewhat aphoristic title of this final chapter of the book, that re-thinking science is not science re-thought. The intention is to stress that there is much work to be done. Though the process of re-thinking science has scarcely begun, some suggestions can be made. To adumbrate these is the function of the seventeen discussion points which follow. In putting them forth, we want to set a small example of what moving into the *agora* might mean.

Debating Points for the *Agora*

Living in a Mode-2 Society

1 Uncertainties, rather than being progressively eradicated, will proliferate

Uncertainties are generated by two forces – the historically unprecedented success of science in producing novelty, and the insatiable demand from society for innovation. Neither force will diminish, but rather they will grow more insistent. Looking back at the twentieth century, which was impregnated by too many – false – certainties, political (including totalitarian) blueprints, economic monopolies and the sometimes haughty aloofness of scientists, we might welcome, not fear, uncertainties that come with innovation, while trying to assure that equalities of opportunity and democratic choices of people are safeguarded and further developed.

2 The challenge now is how best to cope with these uncertainties

Coping with uncertainties has two dimensions. The first is the demand for new social – and not only scientific and technical – innovations to enable individuals and groups to cope with ever encroaching uncertainties. What forms these social innovations might take can only be glimpsed. But there may be clues to be discovered both from how scientists themselves cope with uncertainty (which is an inevitable outcome of the provisionality of scientific results) and also from how, in the past, social mechanisms have been created to cope with uncertainties. We take each in turn. It is difficult not to be struck by the contrast between how uncertainties are perceived in the scientific and the societal domain. In the former, uncertainties are easily accommodated as the necessary (and, indeed, benign) outcomes of a progressive science. Science is, in the words of David Mermin, both outrageously radical and profoundly conservative, meaning that every radically new point of view has to be subtly accommodated into what has already been accomplished, since the valid features of the old perspective must be recognizable within the new framework. What reconciles scientists to endemic uncertainty is the wider sense of security provided by their belief in a progressive science. Indeed, scientists may be the only ones left – or the first ones yet – to uphold a belief in progress. In the societal domain it is different. Uncertainties are threatening because they cannot ultimately be resolved; instead they threaten to spin out of control. But perhaps the crucial difference is this: the consequences of living with uncertainties have to be borne directly by concrete individuals and social groups, while scientific uncertainties are absorbed by a well-institutionalized and protective shield of '*la science c'est nous*'. Two questions arise from these contrasting perceptions of uncertainty as healthy and as destabilizing.

- Is there anything in the training, mentality or culture of scientists (who are able to accommodate uncertainties without feeling threatened) that can be applied to the wider, and more difficult, task of coping with societal uncertainties? Our answer is yes, but only in the limited sense that uncertainties appear less threatening – and can be soothed, although never reduced – within an open environment (such as, generally, prevails in scientific communities where findings are open to interrogation, contradiction and improvement). So it follows that open societies are better able than closed societies to cope with the accumulation of uncertainties. But the main reason for different perceptions of the tractability of scientific and societal uncertainties is that the latter are far more radical, open-ended and de-stabilizing for individuals than the former.

- What strategies have been adopted by individuals and groups to cope with uncertainties – and how successful have they been? One has been to 'spread' these uncertainties, both spatially across personal, social and professional lives, producing and consuming selves, and temporally across life-spans. The well-known and proven mechanisms have been different forms of solidarity, from close kin to the assurance system, all strongly dependent on social ties and economic relations in different configurations. But in an age of increasing individualization – which comes with a decline of traditional forms of solidarity – coping strategies, while still relying on being spread, tend to become ever more internalized. The declining importance of formal work in defining self-worth is due not only to enforced 'external' circumstances such as higher levels of unemployment and the de-professionalization (if not de-skilling) of many occupations, but also to voluntary 'internal' coping strategies, prominent among which is this 'spreading' of uncertainties. Perhaps the spread of so-called portfolio life-styles, or portfolio biographies, is an example.

The adoption of different coping strategies is consistent with the historical evidence provided by social mechanisms that have been adopted in the past to cope with uncertainties. To accommodate, and hopefully reduce, the uncertainties created by the industrial revolution and the advance of an urban, and largely secularized, society, welfare states (following many and enduring conflicts) were created in many countries. Their purpose was – and is – to reassure individuals and ensure that they would not be exposed to the full weight of vicissitudes generated by socio-economic change. To the extent that welfare states are now seen as too expensive and, with greater justification, as too rigid to provide such reassurance because of the explosion of uncertainties produced by the emergence of a post-industrial society, the 'spreading' of these uncertainties by individuals and groups may be an alternative strategy. If this is true, the appropriate public policy response could well be to devise strategies that encourage individuals to 'spread' uncertainties across their plural lives, personal and professional, and across their life-spans from extended education (as represented by the expansion of universities) through 'portfolio' (and shorter?) careers to extended retirement. Such strategies might include the validation of alternatives to paid employment and a drive towards more flexible labour markets as well as traditional social-democratic remedies such as better child care, shorter working-weeks and increased opportunities for education and leisure. It is revealing that governments and parties of the Left, struggling to maintain their traditional social-democratic values and policies, have adopted precisely this 'cocktail' of strategies.

But the management of uncertainties, even in a technical and scientific sense, can no longer be left to scientists (or other narrowly defined groups of experts). One reason for this has already been discussed – namely that scientists are used to accommodating different and less threatening forms of uncertainty and, therefore, are probably ill equipped to advise society at large on the best strategies for coping with uncertainties of a much more radical order. However, the strategies which scientists have traditionally adopted to cope with uncertainty are not entirely irrelevant. Within the context of Mode-2 knowledge production they, too, have to cope with a wider range of less stable uncertainties. They, too, are tending to adopt 'spreading' strategies not only to reduce the risks they run but also to provide themselves with the plural identities needed to accommodate proliferating uncertainties. The growth of multi-disciplinary identities, the emergence of 'trading zones', the development of multi-task teams, may all be evidence of such 'spreading' strategies. If this is accepted, the appropriate science policy response may be to devise policies that encourage, rather than constrain, the development of the heterogeneous knowledge production environments which we have labelled Mode 2.

3 The emergence of Mode-2 society raises acute issues of social justice, economic equality and the further democratization of knowledge

Mode-2 society will lead to enhanced differentiation, greater volatility, increased uncertainties – and, inevitably, less equality. This is not a recent phenomenon, as the turbulence created by the industrial revolution shows. But inequality takes on new forms, and is not only accentuated but accelerated by the advance of globalization. It is further compounded by what we have called the new economic rationality and its valuation of the potential, rather than the actual. Nor can it be dismissed as an unavoidable outcome for at least two reasons. First, the ancient and intractable relationship between knowledge and power remains. As a result the persistence of inequality and unequal access to knowledge/power resources cannot be ignored. Second, without broadly based 'ownership' of, or access to, knowledge, the – virtuous – circle of innovations–uncertainties–innovations will be replaced by a vicious circle. The triple democratization of technology, finance and information, which in the view of some underlies much of the current wave of globalization, is painfully short in being accomplished for the overwhelming majority of human beings on this earth. The optimistic view, of course, is that because of the enhanced capacity for self-organization which we have described, any dysfunctional outcomes are likely to be swiftly self-corrected. But such a view is uncannily reminiscent of the inevitable beneficence of the market's 'hidden hand'. Again the key may be a proliferation, and pluralization, of the parameters which measure equality and inequality. As uncertainties accumulate (and inequalities increase?), so the need for such an extension of criteria will become more urgent. In fact there is some evidence that this is happening; the revival of religion in once thoroughly secularized Western societies is suggestive. Individuals are now seeking a much wider range of arenas in which they can achieve self-identity, self-esteem and self-worth. The aim of public policy, to the extent that it has influence in this private (and privatized?) domain, should be to stimulate this proliferation, and pluralization, of measures of self-worth as an antidote to the anti-egalitarian tendencies within Mode-2 society. Access to knowledge is now as important as the availability of material well-being once promised by the welfare state.

The Context Speaks Back – Transforming Science

4 Contextualization means that the (unknowable) implications as well as the (planned or predictable) applications of scientific research have to be embraced

The context of application is confined to intended – or at any rate knowable – consequences. The context of implication goes much wider, embracing the unintended consequences of knowledge production as well. To the limited extent that there have been attempts to recognize the implications of scientific research, these have been confined to forecasting unintended, and negative, consequences. Contextualization at the level of knowable outcomes is also a less radical project in another sense. It arises from efforts to define problems and then assemble the intellectual, human and financial resources needed to solve them; this is not enough in itself to guarantee the reflexivity characteristic of socially distributed knowledge production. Contextualization that attempts to embrace unpredictable and unintended implications, in contrast, demands reflexivity – in the sense that its purpose is to incorporate future potential into the research process from the very beginning. It goes far beyond a 'forward look' or 'technology foresight' exercise. Of course such an approach is difficult to sustain in systematic or methodological terms. In most 'forward look' or 'technology foresight' exercises, experts offer views about the likely impact of already identified social or technical trends and, through techniques such as the Delphi method, reach consensus about future objectives. But the process is not reflexive – in two senses. First, there is no effort to validate the results of a Delphi consensus or any attempt to update or re-assess it. Second, because the scope for systematic reflection is limited, there is no opportunity to loop-back into the research process. From this it follows that there is a great need to develop further ways in which reflexivity can enter into the actual practices of doing research and setting of research priorities, and of how those who will be impacted by the outcome of research can be involved in an anticipated, open and pluralistic way. Given the state of uncertainty which overshadows necessarily any thinking about the future, nevertheless, a more long-term vision must replace the current insistence on short-term, often purely utilitarian outlooks.

5 Strategies for exploring more accurately the implications of knowledge production need to be developed

The process of inventing, especially in technology, has sometimes been compared to the identification (and subsequent prospecting) of so-called Klondike spaces – spaces which delineate a terrain in which the coveted nuggets will be found. Likewise, one might begin looking for areas in a research terrain where significant implications of particular research projects are likely to arise without being pinpointed exactly. In other words we must 'prospect' for these (presently unknowable) implications. This might involve consulting other knowledge producers and users and also wider social constituencies about the implications of particular projects, in order to gather a diversity of perspectives in a form of 'triangulation' survey. In this way research becomes more reflexive, by gathering together a plurality of views about possible implications and looping them back into the process of innovation. Seen in this light 'backward look' exercises are perhaps as important for identifying implications as 'forward look' are presumed to be for identifying applications. In terms of science policy one possible conclusion is that within every research proposal, and project, there should be a deliberate strategy for identifying – and articulating – its context of implication. This can best be achieved by including those likely to be implicated, perhaps unknowingly, as well as the conscious carriers of social knowledge.

6 Contextualization is pervasive – and must be internalized

Arguably the context of application can be articulated and managed through 'external' mechanisms – such as 'forward look' and 'technology foresight' exercises, science and technology parks, and technology transfer or industrial liaison offices within universities. But the context of implication can only be accommodated through a process of internalization. It speaks back along many different and often informal lines that cannot easily be incorporated into existing administrative routines. These lines of communication need to be encouraged and their value recognized institutionally. Entry and exit points must be identified along the many routes which a particular research process may take. This identification cannot be externalized to communications experts but must be internalized by researchers as a core responsibility.

7 Universities will need to be adaptable organizations (and comprehensive institutions?) rather than specialized organizations (or niche players?)

In order to cope successfully with uncertainties, which they themselves are proliferating, and to articulate the context of implications (and take account of this new context in their academic programmes and teaching styles), universities will have to become more flexible. This has two dimensions. The first is that universities may be unable to react rapidly and creatively to future demands if they are constrained within either a historically determined or bureaucratically imposed division of institutional labour. This is contrary to the common – but, in our view, misguided – view that universities must be re-ordered into a hierarchy of institutions with distinctive missions. The second is that universities, as successful generators of uncertainties, may have a peculiar capacity for accommodating these uncertainties (at any rate, in the research domain) without compromising their particular vision of themselves as universalistic and, in some sense, totalizing institutions. They provide sites in which variation – both intellectual and social – is generated. Yet, despite this, they continue to tell convincing 'stories' of coherence.

Contextualization Means People

8 Contexts are made, not given. Rather they emerge, are generated or constructed, either in relation to particular problems for which they are or may become relevant, or in relation to other, already existing contexts

Contexts do not appear all of a sudden, as being drawn from some 'planning authority'. They are open spatially and temporally – to further evolution and to various communities or groups that attempt to shape them or are drawn into them. Contexts function as a resource or support that the environment offers to those who find themselves in it; but in turn, an individual or social group must possess the capabilities to perceive and use it. Thus, contexts tend to become stabilized after a while in what they have to offer and in the ease with which these resources can be used. This holds for everyday life as much as for research. This dual 'ability' inherent in the relationship between research that is conducted within a given context then has become sufficiently 'matched' in a productive way. Contexts are always multiple and underdetermined in terms of what can be extracted from them, given sufficient ingenuity and 'matching' abilities.

9 People enter. Whenever we speak about the contextualization of knowledge, there is some place in this knowledge where people have entered

People may be conceptualized as objects of research or as the subjects that pursue research, for example, other researchers. They may come in as idealized (and often vague) categories, such as the potential 'users' of a particular research activity. But most importantly, they can be conceptualized either as passive objects or as active agents in the definition of research problems or in exploring the context of implication. In either form, this may be a perfectly legitimate form of conceptualizing them, but research questions and answers that follow from this will tend to be different. Also, the policy implications are likely to differ. Giving a place to people in our knowledge refers to the ways that they are being conceptualized, which may – or may not – be supported by various forms of interaction, communication and participatory engagement with 'real people'. The conventional rhetoric used to establish boundaries still distinguishes between 'making' knowledge, the domain of science, and 'using' (or 'abusing') knowledge, which is accepted to be socially constituted. This distinction emphasizes the neutrality (and so the authority) of science – wrongly, in our view, because it also excludes people and denies that contextualization takes place.

10 Contextualization which contains, or is accompanied by, such a 'human element' is more likely to take subjective experience seriously

In some areas of research, for example, in the bio-medical field, this has become an important complementary source (as well as local, alternative or hybrid source) for how knowledge is being produced and enriched. Contextualization then becomes a multi-layered process, which does not necessarily attempt to strip down everything in a reductionist mode to one 'fundamental' level of explanation. Related to this is a growing awareness that the prevalent division of scientific labour should not necessarily entail a categorical division of social responsibilities. In this view, there is no scientific or technical problem that is 'only' technical or scientific, nor do such exclusively contained solutions exist. Rather, taking people seriously – with their subjective experience, views, wishes and desires as well as interests – will inevitably conjure up some, however vague or remote, sense of being engaged in a wider circle of social accountability, if not responsibility.

11 Whenever people have entered the context and however they may have become conceptualized at whatever stage of the research process, researchers are also engaged in utilizing social knowledge

While researchers may do so implicitly or as part of the tacit knowledge they bring to the task, there are ways of making them more aware of what they are actually doing. Social knowledge therefore can provide an important bridge between research carried out in the natural sciences and many technological fields and how this work will affect people in the social world. Crossing this bridge repeatedly should also make it clear to scientists that contextualization is to be embraced rather than feared. Public contestation, controversy and conflict that appear in the wake of contextualization are not to be shunned on grounds of principle. Rather, they are a sign of a healthy body politic and part of the process of democratization that also affects science. Space has to be made for the discussion of what people want, what their needs are and how even contradictory responses and claims can become better incorporated into the interactive dynamics of exchange between knowledge being produced and the social context in which this occurs.

Reliable Knowledge is not Enough

12 The more open and comprehensive the scientific community the more socially robust will be the knowledge it produces. This is contrary to the traditional assumption that there is a strong relationship between the coherence (and, therefore, boundedness?) of a scientific community and the reliability of the knowledge it produces

On the first of these propositions our argument is, first, that although it might have been true that reliable knowledge was best produced by cohesive (and therefore restricted) scientific communities, socially robust knowledge can only be produced by, and within, much more sprawling socio-scientific constituencies with open frontiers; and secondly, that socially robust knowledge is superior to reliable knowledge for two main reasons. The first reason is that it has been subject to more intensive testing and re-testing in many more contexts (which is why it is 'robust'). Although not invariant in a narrow technical sense, it has a more comprehensive applicability because it is more adaptable. In this sense it aspires to, and perhaps achieves, a different kind of – pluralistic – universalism. The second reason is that because of its malleability and connective capability, socially robust knowledge – which builds upon reliable knowledge while going beyond it – is underdetermined compared to reliable knowledge. Its context is not predetermined, but open to ceaseless re-negotiation. Instead of achieving a precarious invariance by establishing strict limits within which its truthfulness can be tested, as reliable knowledge does, socially robust knowledge is the product of intensive (and continuous) interaction between results and their interpretation, people and environments, applications and implications.

13 A variety of knowledge traditions is needed constantly to replenish the epistemological core. Again this is antagonistic to the universalistic claims of Western science, which seeks to exclude the locally contingent in order to produce universal and 'invariant' results

On the second proposition we argue for a greater variety of knowledge traditions alongside the 'mainstream' Western Enlightenment tradition – local (in the sense that they stand outside that tradition), alternative (in the sense of being perhaps even actively opposed to it) or hybrid (in the sense of combining scientific and other knowledge elements, and therefore running the risk perhaps of being labelled 'impure' by traditional scientists). We do so on two main grounds. The first is that what we have called the epistemological core is constantly changing, being emptied and then filled again with new values, norms and practices. It is not fixed, a hard dead core, unchanging and unchangeable. So openness to a greater variety of knowledge traditions is a way constantly of re-activating the creativity of the core. The second is that such is the richness of the many knowledge traditions which have been generated in the past and are continually formed and reformed, the risk of collapsing into relativism is actually slight. These riches allow us constantly to re-configure knowledge. Although it is true always that 'it depends', this does not mean that 'anything goes', and, therefore, there is no point of rigorously pursing one truth; rather, the variety, the comparabilities and incommensurabilities, of these different knowledge traditions encourage us to seek more refined and subtle truths. To assert this is to keep faith with the spirit of the Enlightenment. To overdetermine the paths and outcomes of science is to deny the open-endedness of the Enlightenment project.

14 Images of science, even populist representations, support rather than detract from the body of knowledge produced by research. However, the gap between images of science and the actual practices should not become too wide. This proposition too is contrary to the traditional assumptions that 'stories' about science may encourage the belief that science is merely a set of incommensurable (or relativistic) discourses, and that the popularization of science may encourage unwelcome 'lay' interventions (unless carefully

managed as part of a campaign to increase the public understanding of science)

On this proposition our argument is that in an age of intense contextualization, images of science need to have a strong 'reality-content', that is, be closer to actual practices and their rapid changes than the traditional and timeless images of Science and its Pursuit of Truth suggest. Images too are 'inside' the process of contextualization and get shaped accordingly. They are even 'inside' the core, in the sense that these images enable us to construct a more subtle and sociologically sensitive epistemology; they also represent the soft layers (in Weinberg's phrase) through which the context penetrates the core. We assert, first, that it is no longer possible to establish a clear demarcation between context and core, hard layers and soft layers, the body of knowledge and images of knowledge. Many of the most promising scientific arenas combine elements of both. Secondly, under Mode-2 conditions it is wrong to regard the intrusion of the 'social' (which perhaps sums up context, soft layers and images in a single word) as inhibiting, or even destroying, scientific novelty; instead, it is a key source of creativity and so of innovation.

Science Enters the Agora

15 If the *agora* has become the space in which science meets and interacts with many more agents, where institutions overlap and interact and where interests, values and actual decisions to be taken are being discussed, negotiated, fought over and somehow settled, then the self-organizing capacity of all participants needs to be enhanced

What might be done to achieve this? Here, there is a tension between individual or institutional autonomy and the spread of the imperatives of the so-called audit society. Increasing self-organizing capacity is not simply a matter of increasing the autonomy of participants *vis-à-vis* one another. On the contrary, participants need to be made more responsive to them and to each other; that is, they need to act more reflexively. At the same time, one cannot enhance the responsiveness by simply increasing the demand for public accountability. Externally applied, this could move participants into more, rather than less, defensive positions. A possible way forward, which might also form an indicator of increasing self-organization, would be the extent to which participants are internalizing accountability of their own volition. Thus, to enhance self-organization would presuppose establishing the conditions in which participants are willing to behave in this way. It might not be too far-fetched to suggest that there is a relation of analogy between autonomy and self-organization on the one hand and reliable knowledge and socially robust knowledge on the other. In the *agora* the conditions which promote more self-organization also promote the generation of socially robust knowledge.

16 If expertise is becoming socially widely distributed and transgressive, trust becomes an even more scarce and precious resource

Traditionally, trust has been established in three main ways: (1) deference to an already established authority which enjoys a high reputation and the advantages that come from drawing on it; (2) face-to-face interaction which presupposes prolonged interaction among members who know each other and who can and will retaliate, if trust and the expectation of reciprocity are broken; and (3) reliance on 'mechanical objectivity', that is, previously agreed-upon means of how to arrive at and interpret, and which conclusions to draw from, 'externally' arrived at evidence. While all three of these traditional sources of trust continue to operate, they are expanded, stretched, tested and re-interpreted – or even re-invented – under a regime of socially distributed expertise. None of them can be taken for granted any more; deference to scientific authority has been on the wane for some time, and reputation, while continuing to be crucial, is made and re-made in accordance with criteria whose exact composition is bound to vary. Thus, for instance, the media are playing a much more important part in the making and re-making of scientific authority and the credibility that goes with it. Face-to-face contact, even within the relatively narrow confines of a scientific community, has equally become diminished, while new opportunities for functional equivalents in a globalizing world open up through substitutes like those offered by the Internet and other forms of 'virtual' communication. The latter introduce new elements of volatility and transience. 'Mechanical objectivity' as a source of trust has come under attack, whenever the consensus that underpins it is itself open to critical debate or controversy.

We expect that trust, in order to be maintained or to be re-created, will have to partly re-balance its ingredients and the basis from which it draws its authority, as well as having to follow more 'local' patterns of establishing and maintaining itself. It cannot be taken for granted among the continuing flow of uncertainties that are intrinsically connected with the innovative capacities of science. 'Believing' and knowing whom to believe, when and to what extent, as well as in what sense, becomes part of 'knowing'. It no longer follows automatically from 'belonging' to either an expert or a lay community. It is in this sense that 'we have all become experts now', but this expertise, widely distributed, is continuously being tested, as well as contested, before being trusted.

17 The call for more 'participation', epitomized in an imperative like 'participate or perish', is not to be taken as a free entry ticket into an inchoate and unstructured arena of endless (and often futile) debates

Just as 'publish or perish' is underpinned by certain rules of the game, to which scientists and their peers have agreed to adhere, so the opening up of science towards the *agora* presupposes and necessitates 'rules' of a game that partly still wait to be established. When modern science became institutionalized, one of its much envied strengths was its ability to create consensus. This was partly achieved through mechanisms of exclusion – of certain themes, for instance – and partly through mechanisms of inclusion – by admitting (non-scientific, but credible) witnesses as additional (social) sources for establishing what counted as a scientific 'fact'. If the *agora* calls for a widened notion of thus established reliable knowledge, by making it more socially robust, the rules to achieve this partly still need to be defined and agreed upon. Not everyone will be able or willing to participate, and not anything goes – but the often feared 'contamination' of science through the social world should be turned around. Science can and will become enriched by taking in the social knowledge it needs in order to continue its stupendous efficiency in enlarging our understanding of the world and of changing it. This time, the world is no longer mainly defined in terms of its 'natural' reality, but includes the social realities that shape and are being shaped by science.

References

Adams, J., Bailey, T., Jackson, L., Scott, P., Pendlebury, D. and Small, H. 1998: *Benchmarking of the International Standing of Research in England*. Leeds: Centre for Policy Studies in Education, and Philadelphia: Institute for Scientific Information.

Annales 1989: Tentons l'expérience. *Annales ESC*, 44, 1317–23.

Appiah, Kwame A. 1998: Africa: The hidden history. *New York Review of Books*, 17 December, 64–72.

Balmer, Brian 1996: Managing mapping in the Human Genome Project. *Social Studies of Science*, 26, 531–73.

Beck, Ulrich 1992: *Risk Society: Towards a New Modernity*. London: Sage.

Bell, Daniel 1973: *The Coming of Post-Industrial Society*. London: Heinemann.

Bell, Daniel 1976: *The Cultural Contradictions of Capitalism*. London: Heinemann.

Bernard, Claude 1865/1966: *Introduction: l'étude de la médecine expérimentale*, ed. François Dagognet. Paris: Garnier-Flammarion.

Blume, Stuart 1987: The theoretical significance of cooperative research. In Stuart Blume, Joske Bunders, Loet Leydesdorff and Richard Whitley (eds), *The Social Direction of the Public Sciences. Sociology of the Sciences, A Yearbook*, 11, Dordrecht: Reidel, 3–38.

Boyle, Nicholas 1991: *Goethe: The Poet and the Age*, vol. 1: *The Poetry of Desire*. Oxford: Clarendon Press.

Callon, Michel 1989: *La Science et ses réseaux. Genèse et circulation des faits scientifiques*. Paris: Editions de la Découverte.

Carr, Geoffrey 1998: A survey of the pharmaceutical industry. *The Economist*, 346, after p. 72.

Cartwright, Nancy, Cat, Jordi, Fleck, Lola and Uebel, Thomas 1996: *Otto Neurath: Philosophy between Science and Politics*. Cambridge: Cambridge University Press.

Clancy, Carolyn and Eisenberg, John 1998: Outcomes research: Measuring the end results of health care. *Science*, 281, 245–6.

Coleman, James 1987: Norms as social capital. In Gerard Radnitzky and Peter Bernholz (eds), *Economic Imperialism*, New York: Paragon House, 133–55.

Collins, F., Patrinos, A., Jordan, E., Chakravati, A., Gesteland, R. and Walters, L. 1998: New goals for the U.S. Human Genome Project: 1998–2003. *Science*, 282, 882–9.

Darnton, Robert 1999: The new age of the book. *New York Review of Books*, 18 March, 57.

Daston, Lorraine 1998: Fear and loathing of the imagination in science. *Daedalus*, 127, 73–95.

Daston, Lorraine and Galison, Peter 1992: The image of objectivity. *Representations*, 40 (Fall), 81–128.

David, Paul 1995: Science reorganized? Post-modern visions of research and the curse of success. Ms. December 1995.

Deuten, Jasper, Rip, Arie and Jelsma, Jaap 1997: Societal embedment and product creation management. *Technology Analysis and Strategic Management*, 9, 219–36.

Dodgson, Mark 1993: *Technological Collaboration in Industry: Strategy, Policy and International Innovation*, London: Routledge.

Dosi, Giovanni 1996: The contribution of economic theory to the understanding of a knowledge-based economy. In *Employment and Growth in the Knowledge-based Economy*, Paris: OECD, 81–92.

Douglas, Mary 1986: *How Institutions Think*. Syracuse, NY: Syracuse University Press.

Elias, Norbert 1937/1982: *The Civilizing Process: The History of Manners and State Formation and Civilization*. Oxford: Blackwell.

Elkana, Yehuda 1981: A programmatic attempt at an anthropology of knowledge. In Everett Mendelsohn and Yehuda Elkana (eds), *Sciences and Cultures. Sociology of the Sciences, A Yearbook*, 5, Dordrecht: Reidel, 1–76.

Epstein, Steven 1997: *Impure Science: AIDS, Activism, and the Politics of Knowledge*. Berkeley: University of California Press.

Ewald, François 1986: *L'Etat providence*. Paris: Grasset.

Ezrahi, Yaron 1990: *The Descent of Icarus: Science and the Transformation of Contemporary Democracy*. Cambridge, MA: Harvard University Press.

Felt, Ulrike 1997: *Wissenschaft auf der Bühne der Öffentlichkeit. Die alltägliche Popularisierung der Naturwissenschaften in Wien, 1900–1938*. Wien: Habilitationsschrift an der Grund- und Intregrativwissenschaftlichen Fakultät der Universität Wien.

Fleck, Ludwik 1935/1979: *Genesis and Development of a Scientific Fact*. Chicago: University of Chicago Press.

Friedman, Thomas 1999: *The Lexus and the Olive Tree*. London: Harper Collins.
Fuller, Steve 1997: *Science*. Buckingham: Open University Press.
Gago, José M. 1998: *The Social Sciences Bridge*. Lisbon: Observatorio das Ciencias e das Tecnologias.
Galison, Peter 1997: *Image and Logic: A Material Culture of Microphysics*. Chicago/London: University of Chicago Press.
Galison, Peter and Stump, David J. (eds) 1996: *The Disunity of Science: Boundaries, Contexts, and Power*. Palo Alto, CA: Stanford University Press.
Georghiou, Luke 1998: Science, technology and innovation policy in the 21st century. *Science and Public Policy*, 25, 1–3.
Gibbons, Michael, Limoges, Camille, Nowotny, Helga, Schwartzman, Simon, Scott, Peter and Trow, Martin 1994: *The New Production of Knowledge: The Dynamics of Science and Research in Contemporary Societies*. London: Sage.
Giddens, Anthony 1984: *The Constitution of Society: Outline of the Theory of Structuration*. Cambridge: Polity Press.
Giddens, Anthony 1992: *The Transformation of Intimacy*. Cambridge: Polity Press.
Gieryn, Thomas F. 1999: *Cultural Boundaries of Science: Credibility on the Line*. Chicago: University of Chicago Press.
Gizycki, Rainald von 1987: Cooperation between medical researchers and a self-help movement: The case of the German Retinitis Pigmentosa Society. In Stuart Blume, Joske Bunders, Loet Leydesdorff and Richard Whitley (eds), *The Social Direction of the Public Sciences. Sociology of the Sciences, A Yearbook*, 11, Dordrecht: Reidel, 75–88.
Goldemberg, José 1998: What is the role of science in developing countries? *Science*, 279, 1140–1.
Goldhagen, Daniel 1996: *Hitler's Willing Executioners*. London: Little, Brown.
Grafton, Anthony 1999: Remaking the Renaissance. *New York Review of Books*, 4 March, 34–8.
Graham, Loren 1998: *What Have We Learned About Science and Technology from the Russian Experience?* Palo Alto, CA: Stanford University Press.
Greif, Avner 1997: Cultural beliefs as a common resource in an integrating world. In Partha Dasgupta, Karl-Goeran Maeler and Alessandro Vercelli (eds), *The Economics of Transnational Commons*, Oxford: Clarendon Press, 238–96.
Hacking, Ian 1999: *The Social Construction of What?* Cambridge, MA: Harvard University Press.
Haldane, J. B. S. 1923/1995: Daedalus or science and the future. A paper read to the heretics, Cambridge, on February 4, 1923. In Krishna R. Dronamraju (ed.), *Haldane's Daedalus Revisited*, Oxford/New York/ Tokyo: Oxford University Press, 23–54.

Hart, David M. and Victor, David G. 1993: Scientific elites and the making of US policy for climate change research, 1957–74. *Social Studies of Science*, 23, 643–80.

Herbert, Ulrich 1998: Vernichtungspolitik. Neue Antworten und Fragen zur Geschichte des Holocaust. In Ulrich Herbert (ed.), *Nationalsozialistische Vernichtungspolitik 1939–1945*, Frankfurt a. M.: Insel Taschenbuch Verlag, 9–66.

Hicks, Diana and Katz, J. Sylvan 1996: Science policy for a highly collaborative science system. *Science and Public Policy*, 23, 39–44.

Hirschman, Albert O. 1970: *Exit, Voice and Loyalty. Responses to Decline in Firms, Organizations, and States*. Cambridge, MA: Harvard University Press.

Hughes, Thomas P. 1983: *Networks of Power*. Baltimore/London: Johns Hopkins University Press.

Hughes, Thomas P. 1987: The evolution of large technological systems. In Wiebe E. Bijker, Thomas P. Hughes and Treor Pinch (eds), *The Social Construction of Technological Systems*, Cambridge, MA: MIT Press, 51–82.

Hughes, Thomas P. 1998: *Rescuing Prometheus*. New York: Pantheon Books.

Irwin, Alan 1995: *Citizen Science: A Study of People, Expertise and Sustainable Development*. London: Routledge.

Jacob, François 1981: *Le Jeu des possibles: Essai sur la diversité du vivant*. Paris: Fayard.

Jasanoff, Sheila 1997: NGOs and the environment: From knowledge to action. *Third World Quarterly*, 18, 579–94.

Kaku, Michio 1998: *Visions: How Science Will Revolutionize the Twenty-First Century*. Oxford: Oxford University Press.

Kennedy, P. 1993: *Preparing for the Twenty-First Century*. New York: Random House.

Kevles, Daniel 1987: *The Physicists: The History of a Scientific Community in Modern America*. Cambridge, MA: Harvard University Press.

Keynes, John Maynard 1937: The general theory of employment. *Quarterly Journal of Economics*, 51, 209–23.

Kleiber, Charles 1999: *Die Universität von morgen: Visionen, Fakten, Einschaetzungen/Pour l'université: histoire, état des lieux et enjeux: l'université de demain: opinions et debats*. Bern: Gruppe für Wissenschaft und Forschung.

Knorr Cetina, Karin 1999: *Epistemic Cultures: How the Sciences Make Knowledge*. Cambridge, MA: Harvard University Press.

Krige, John (ed.) 1996: *History of CERN*. 3 vols, Amsterdam: Elsevier.

Krige, John and Pestre, Dominique (eds) 1997: *Science in the Twentieth Century*. Amsterdam: Harwood Academic Publishers.

Kruckemeyer, Kenneth 1998: Managing public participation in public works projects *Independent Activities Program Course*, MIT, January, 1995, notes taken by and quoted from Loren Graham, *What Have We Learned*

About Science and Technology from the Russian Experience? Palo Alto, CA: Stanford University Press.
Küppers, G., Lundreen, P. and Weingart, P. 1978: *Unweltforschung – Die Gesteuerte Wissenschaft?* Frankfurt a.M.: Campus.
Latour, Bruno 1997: Socrates' and Callicles' settlement – or, the invention of the impossible body politic. *Configurations*, 5, 189–240.
Latour, Bruno 1998: From the world of science to the world of research? *Science*, 280, 208–9.
Luhmann, Niklas 1996: Modern society shocked by its risks. Social Sciences Research Centre, Occasional Paper 17, Hong Kong, 3–19.
MacKenzie, Donald 1996: How do we know the properties of artefacts? Applying the sociology of knowledge to technology. In Robert Fox (ed.), *Technological Change: Methods and Themes in the History of Technology*, Amsterdam: Harwood Academic Publishers, 247–63.
Markus, Gyorgy 1987: Why is there no hermeneutics of natural sciences? Some preliminary theses. *Science in Context*, 1, 50–1.
Mermin, David 1998: The science of science: A physicist reads Barnes, Bloor and Henry. *Social Studies of Science*, 28, 603–23.
Merton, Robert K. 1942: Science and technology in a democratic order. *Journal of Legal and Political Sociology*, 1, 115–26.
Nature 1999: Dangers of Euro-relevance. *Nature*, 398, 1.
Nelkin, Dorothy 1987: *Selling Science: How the Press Covers Science and Technology*. New York: Freeman.
Nowotny, Helga 1990: Actor-networks vs. science as self-organizing system: A comparative view of two constructivist approaches. In Wolfgang Krohn, Günter Küppers and Helga Nowotny (eds), *Self-Organization – Portrait of a Scientific Revolution. Sociology of the Sciences, A Yearbook*, 14, Dordrecht: Kluwer, 223–39.
Nowotny, Helga 1994: *Time: The Modern and Postmodern Experience*. Cambridge: Polity Press.
Nowotny, Helga 1999: The place of people in our knowledge. *European Review*, 7, 247–62.
Nowotny, Helga 2000: Transgressive competence: The changing narrative of expertise. *European Journal of Social Theory*, 3(1), 5–21.
Nowotny, Helga and Felt, Ulrike 1997: *After the Breakthrough: The Emergence of High-Temperature Superconductivity as a Research Field*. Cambridge: Cambridge University Press.
Nussbaum, Martha 1997: *Cultivating Humanity: A Classical Defense of Reform in Liberal Education*. Cambridge, MA: Harvard University Press.
Pestre, Dominique 1999: What is science? A simple question with no simple answer. In Gerd Folkers, Helga Nowotny and Martina Weiss (eds), *Sternwarten–Buch: Jahrbuch des Collegium Helveticum* 2, Zürich: Haffmans Sachbuch Verlag, 161–73.
Popper, Karl 1969: *Conjectures and Refutations: The Growth of Scientific Knowledge*. London: Routledge & Kegan Paul.

Porter, Theodore M. 1995: *Trust in Numbers: The Pursuit of Objectivity in Science and Public Life*. Princeton, NJ: Princeton University Press.

Postrel, Virginia 1999: *The Future and its Enemies: The Growing Conflict Over Creativity, Enterprise and Progress*. New York: The Free Press.

Power, Michael 1997: *The Audit Society: Rituals of Verification*. Oxford: Oxford University Press.

Rabinow, Paul 1992: Artificiality and Enlightenment: From sociobiology to biosociality. In Jonathan Crary and Sanford Kwinter (eds), *Incorporations*, New York: Zone, 234–52.

Rapp, Rayna, Heath, Deborah and Taussig, Karen Sue 1998: Tracking the Human Genome Project. *Items*, 52, 88–91.

Reich, R. 1992: *The Work of Nations*. New York: Vintage Books.

Rheinberger, Hans-Jörg 1994: Experimental systems: Historiality, deconstructions, and the 'epistemic thing'. *Science in Context*, 7, 65–81.

Rip, Arie, Misa, Thomas and Schot, Johan 1995: Constructive technology assessment: A new paradigm for managing technology in society *and* Epilogue. In Arie Rip, Thomas Misa and Johan Schot (eds), *Managing Technology in Society: The Approach of Constructive Technology Assessment*, London: Pinter, 1–12 and 347–54.

Robbins-Roth, Cynthia (ed.) 1998: *Alternative Careers in Science: Leaving the Ivory Tower.* San Diego, CA: Academic Press.

Roqueplo, Phillippe 1997: *Entre Savoir et Décision. L'expertise scientifique*. Paris: INRA.

Rorty, Richard 1991: Solidarity or objectivity? In *Objectivity, Relativism, and Truth*. Cambridge: Cambridge University Press, 21–34.

Runciman, W. G. 1966: *Relative Deprivation and Social Justice: A Study of Attitudes to Social Inequality in Twentieth-Century England*. Berkeley: University of California Press.

Schaffer, Simon 1997: What is science? In John Krige and Dominique Pestre (eds), *Science in the Twentieth Century*, Amsterdam: Harwood Academic Publishers, 27–41.

Schatz, Gottfried 1998: The Swiss vote on gene technology. *Science*, 281, 1810–11.

Schon, Donald A. 1967: *Technology and Change: The New Heraclitus*. Oxford: Pergamon Press.

Scott, Peter 1998: The end of the European university. *European Review*, 6, 455.

Sen, Amartya 1999: *Development as Freedom*. Oxford: Oxford University Press.

Sennett, Richard 1998: *The Corrosion of Character: The Personal Consequences of Work in the New Capitalism*. NewYork/London: W. W. Norton & Company.

Star, Susan Leigh and Griesemer, James R. 1989: Institutional ecology, 'translations' and boundary objects: Amateurs and professionals in Berkeley's Museum of Vertebrate Zoology, 1907–39. *Social Studies of Science*, 19, 387–420.

Stehr, Nico 1994: *Arbeit, Eigentum und Wissen: Zur Theorie von Wissensgesellschaft*. Frankfurt a.M.: Suhrkamp.

Strathern, Marilyn 1996: *A Case of Self-Organisation*. Ms. 1996.

Strathern, Marilyn 2000: The tyranny of transparency. In Helga Nowotny and Martina Weiss (eds), *Shifting Boundaries of the Real: Making the Invisible Visible,* Zürich: vdf, 59–78. Reprinted in *British Educational Research Journal*, 26, 309–21.

Traweek, Sharon 1988: *Beamtimes and Lifetimes: The World of High-Energy Physicists*. Cambridge, MA/London: Harvard University Press.

Turney, Jon 1998: *Frankenstein's Footsteps: Science, Genetics and Popular Culture*. New Haven: Yale University Press.

Utterback, J. 1994: *Mastering the Dynamics of Innovation*. Cambridge, MA: Harvard University Press.

van den Daele, Wolfgang 1996: Objektives Wissen als politische Ressource. Experten und Gegenexperten im Diskurs. In Wolfgang van den Daele and Friedrich Neidhardt (eds), *Kommunikation und Entscheidung*, Berlin: Edition Sigma, 297–326.

van Duinen, Reinder J. 1998: European research councils and the Triple Helix. *Science and Public Policy*, 25, 381–6.

Vaughan, Diane 1996: *The Challenger Launch Decision*. Chicago: University of Chicago Press.

Weinberg, Alvin 1963: Criteria for scientific choice. *Minerva*, 1, 159-71.

Weinberg, Steven 1998: The revolution that didn't happen. *New York Review of Books,* 8 October, 48–52.

Weingart, Peter 1997: From finalisation to Mode 2: Old wine in new bottles? *Social Science Information*, 36, 4.

Wennerås, Christine and Wold, Agnes 1997: Nepotism and sexism in peer-review. *Nature*, 387, 341–3.

Whitley, Richard 1999: The institutional structuring of innovation strategies and technological development. Ms. prepared for the Sociology of Sciences Meeting, Uppsala, Sweden, September 1999.

Wilson, Edward O. 1998: *Consilience: The Unity of Knowledge*. New York: Knopf.

Wynne, Brian 1996: Misunderstood misunderstandings: Social identities and public uptake of science. In Alan Irwin and Brian Wynne (eds), *Misunderstanding Science? The Public Reconstruction of Science and Technology,* Cambridge: Cambridge University Press, 19–46.

Yearley, S. 1996: *Sociology, Environment, Globalisation*. London: Sage.

Ziman, John 1991: *Reliable Knowledge: An Exploration of the Grounds for Belief in Science*. Cambridge: Cambridge University Press.

Ziman, John 1996: Is science losing its objectivity? *Nature*, 382, 751–4.

Ziman, John 1998: Why must scientists become more ethically sensitive than they used to be? *Science*, 282, 1813–14.

Ziman, John 2000: *Real Science: What It Is, and What It Means*. Cambridge: Cambridge University Press.

Index

academic ethos 46, 54, 174, 177
academies of arts and sciences 55
Academy of Sciences 85
accelerators 57, 126
accountability 19, 24, 34, 45, 47, 60–1, 68, 73, 83, 87, 119, 126–7, 204, 210–11, 220, 235, 239, 257, 260-1
activist organizations 74, 83, 133, 211
Advanced Research Projects Agency Network (ARPANET) 135
Advisory Council for Science and Technology (ACOST) 149
after 1945 5–6, 60, 70, 81, 122, 204, 206; *see also* post-war period
Africa 162
agora 23, 55, 92, 177, 183–4, 197–9, 201–15, 227–8, 230, 233, 240, 243, 247–9, 260–2
AIDS 115, 211
alternative medicine 36
American Academy of Sciences 235
American Physical Society 213
animal experiments 110
Annales 187
anthropology 125, 145, 160, 162, 188, 229
applicability 56
applied science 4, 38, 53, 113, 122, 155, 184, 217, 220
arms race 107
Arnold, Matthew 81
art 12, 27–8, 242–3
astrophysics 41
audit 45, 83, 114–15, 235, 238–9, 260
Audit Society, The 45
Australia 85
Austria 85

Bacon, Francis 31, 122, 199
Balmer, Brian 148–9
basic research 11, 38, 47, 53, 67, 70, 76, 128, 155–7
Bayh-Dole Act 60
Beck, Ulrich 14, 17, 27, 35, 44
behavioural beliefs 101
belief systems 106, 108–10, 131, 146, 158, 160, 168, 232
Bell, Daniel 12–14
Bernard, Claude 61
Big Science 57, 156
Bildung 81
Bildungsbürgertum 27
binary systems of higher education 85–7
biological sciences 72, 126, 140–1
bio-medicine 36, 40, 61, 207, 257
biotechnology 58, 60, 183, 209, 231
black-boxing 136
blind variation 112, 114–15, 133, 184
Blume, Stuart 139
Bodmer, Walter 149

boundary objects 29, 148
boundary work 57
Bourdieu, Pierre 188
Boyle, Nicholas 27
Brenner, Sidney 149
Britain 86–7, 181
  see also Scotland; United Kingdom
Brookhaven 124
Brundlandt Commission 7
BSE 223
business 14, 24, 28, 46, 49, 60, 67, 76, 88, 101, 137, 139, 153, 170–1, 181–2, 207, 212, 221

California master plan 85
capitalism 5, 8–9, 32–3, 203
career 33, 82–3, 100, 202
Central Artery/Tunnel Project 121, 134–9, 142
CERN 100, 124
certainty trough 42
chaos theory 5, 7–8
chemistry 54
Chernobyl 34, 208
China 138
church 55
citizens 15, 24, 40, 51, 57, 110, 117–19, 137, 192, 203, 206, 214, 220, 226
civil servants 49, 80, 130, 149, 218–19
clinical trials 72, 131, 211, 213
cloning 14, 36
CNRS 85–7
co-evolution of science and society 30–1, 33, 35, 40–1, 43, 46, 48–9, 66–7, 112, 179, 204, 243, 245, 248
cognitive authority 55, 179, 184–8, 191, 196, 200, 207, 211, 230, 233–4
Cold War 6, 8–9, 13–14, 67, 107, 124, 126–7
collaboration 112–13, 125, 145, 242
collectivism 102–3, 108, 110, 125, 228
commodification 21
communication 51, 80, 110, 131–2, 134, 140–1, 143, 146–7, 152, 170, 175
Communism 8, 12, 32, 107
communitarian regime 97, 125, 190
competition and competitiveness 13, 31, 35, 53, 61, 67–8, 72, 102, 107, 110, 112, 115–16, 129, 145, 151–3, 156, 158, 180
complexity 4, 30, 40, 44–5, 47, 77, 86, 113, 133, 135, 139, 158, 202, 206, 222, 225, 245, 247–8
consensibility 170–6, 199
consensuality 170–7, 199
consensus conferences 34, 133
context of application 1, 17, 20, 22, 29, 69, 97, 112, 144, 157, 159, 161, 174, 201, 206, 223, 232–3, 240, 253, 255
context of implication 17, 20, 29, 97, 144, 158–65, 201, 232–3, 240, 253, 255–6
context-sensitivity 117–18, 130, 135
contextualization 1, 4, 17–19, 27, 49–50, 52–6, 58, 61–2, 64–6, 69, 71–4, 77, 80, 84, 89–90, 94–7, 100, 102, 106–11, 113–15, 119–21, 123, 127–8, 134, 138–9, 143–4, 147, 149, 151, 154–5, 157, 160, 165–8, 179, 191, 194–6, 199, 201, 205–11, 230, 235, 240, 242, 245–8, 253–8, 260
contextualization of the middle range 97, 120, 143–65
Co-operative Research Network 242
corporate laboratories 69
corporate universities 15, 92
cosmology 41
cost effectiveness 75
counter-culture movement 136, 138
credentialization 83
credit transfer 85
cultural industries 9, 16, 26–8, 166
culture 1–2, 4, 12–13, 16, 18, 21–2, 26–30, 47, 67–8, 80, 108, 110, 121, 152, 166, 179, 186–7, 189, 206, 243–5
culture wars 81

curiosity-driven science 78, 122
curriculum 82, 91, 94, 145, 212, 242
customization 11, 40, 205–6

Darwin, Charles 184, 189, 195
Darwin, Erasmus 189
Daston, Lorraine 64, 169
David, Paul 4
decentralization 7, 41, 71
decision-making 34, 119, 123–4, 133, 137, 153, 202, 212, 216–18, 220–3, 226
de-differentiation 29, 38, 67, 94
deference 5, 81, 84, 136, 166, 227, 261
de-institutionalization 43, 91–3
demarcation criteria 56–7
democracy 14, 45, 81, 83, 94, 169, 204, 210, 233–4, 240
Denkkollektiv 62, 234
de-regulation 32, 192
design 107, 127, 135, 137–8, 144–5, 150, 205–6
design configuration 153–4
de-synchronization 40
determinism 5, 8, 12, 30
Dewey, John 51
DFG (Deutsche Forschungsgemeinschaft) 59
directed research programmes 73, 76–8
disciplinary structure 84, 105–6, 111, 113, 145, 155, 168, 173, 176, 220
disinterestedness 55, 58, 174
dissemination of research findings 74
dissonance 40, 56
diversification 104
diversity 21, 42, 56, 68, 82, 85, 91, 97, 111–12, 180
'Dolly' the sheep 36, 183
Douglas, Mary 45
dual-support system 86–7
Durkheim, Emile 218

economic growth 6–7, 9, 40, 107, 151, 181, 240
economic rationality 7, 37–8, 48, 237–8, 245, 252
economics 38, 40, 162
education 11, 74, 81, 85, 92, 117
Elias, Norbert 45
Eliot, George 138
elite universities 81–3, 85, 88–90, 95
Elkana, Yehuda 193, 195
employment 11–12, 23, 32, 43, 83, 118, 152
energy production 105
England *see* Britain
Enlightenment 13, 41, 63, 181, 189, 191, 204, 234, 240, 259
enterprise state 12
entertainment 32
entrepreneurial universities 92
entrepreneurship 58, 109, 132, 145
environmental concerns 2, 7, 13, 128, 132–3, 137, 139, 152, 166, 202, 213, 217
epistemic culture 97, 100, 125
epistemological core 19, 54, 58, 64, 66, 93–4, 165, 179, 185–8, 192, 196, 199, 201, 233, 246, 259
Ethical Legal Social Implications (ELSI) programme 239
ethics committees 202, 213, 235, 239
European Commission 77, 129, 223
European Union 22, 53, 67, 70, 75–8, 241
experimentation 112–13, 116, 133, 145, 165, 188
experts 5, 10, 19, 35, 42–3, 54–5, 64, 76–7, 80, 83–4, 88, 99, 105, 108, 110, 113, 116, 136, 142, 150, 171, 173, 186, 188, 198, 208, 210, 215–16, 218–20, 222–3
expertise 16, 19, 105–6, 108, 140, 153, 166, 176, 188, 214–29, 247, 261
Ezrahi, Yaron 23

Fachhochschulen 85–6
family 11–12, 14, 26, 140, 160–1, 191
Felt, Ulrike 155

feminist thinking 60, 116, 140, 212
Feyerabend, Paul 56
Fifth Framework Programme 53, 70, 75, 77, 129, 241
finance 67, 74, 106, 109, 127, 152, 202, 218, 253
Fleck, Ludwik 62, 234
focus groups 34
foundations 128
Fourth Framework Programme 53
France 86, 140, 219
Frankenstein 36
Fukuyama, Francis 14
Fuller, Steve 185–7
funding councils 56, 87, 158, 207

Galileo 122, 206
Galison, Peter 64, 103, 143–7, 169, 175
gender 11, 14, 60, 116, 212–13
gene sequencing 14
genetically modified plants 208
Germany 27, 59, 81, 85–6, 108, 162–3, 181, 208
Giddens, Anthony 15, 17, 19, 28, 44
global brands 16, 22, 25
global warming 13
globalization 7–8, 13, 17, 22–3, 31, 35, 41, 52, 75, 156, 180–1, 229, 252–3
Goethe 27
Goldhagen, Daniel 163
gossip circles 125
government 5, 16, 23, 32, 53, 69, 73–8, 88, 95, 107, 123–4, 128–30, 132, 135, 141, 149, 151–2, 174, 208, 210, 213, 223, 233, 241, 247, 252
government research establishments 50, 60, 69–71, 73, 96, 166, 175
graduates 74, 83–4, 90, 105, 109, 204
Graham, Loren 138
grandes écoles 86, 219
Greece 203, 242
Greif, Avner 101–2, 107, 124

Hamburg 124
HBO (higher professional schools) 85
health and safety 71, 152, 154, 212
health service 24, 40, 206
heterogeneity 8, 19, 28, 42, 112, 136, 144, 223, 229
Hicks, Diana 3
high energy physics 97–100, 103, 114, 194
high temperature superconductivity (HTS) 155–8
higher education 3, 9, 24, 49, 80–1, 83–8, 93–4, 105, 193, 218; *see also* university
Higher Education Funding Council for England 87
Hirschman, Albert 203
history 101–2, 108, 122, 124, 135, 146, 159, 161–5, 181, 188–9, 204
Hitler's Willing Executioners 163
Hobsbawm, Eric 231
Holocaust 162–4
Homer 27
homogeneity 8, 42, 112, 211
Hughes, Thomas 135–6
Human Genetics Advisory Commission 224
Human Genome Mapping Project (HGMP) (UK) 148–51
Human Genome Project (US) 159–61, 183, 239
humanities 3, 72, 81, 106, 117, 164, 189
Huxley, Thomas Henry 81
hybridization 106
hypersonic aircraft 150–1

incommensurability thesis 146, 192, 195
independent research institutions 85–8
individualization 11, 13, 17, 22, 205, 251
industrial laboratories 50, 59, 70, 122
industrial restructuring 12, 22, 71
industrial revolution 243, 251–2

industrial society 14, 29, 68, 105, 108, 111, 124, 205
information and communication technologies 7, 10, 22, 29, 32–3, 41, 67–8, 92, 105, 129, 135, 185, 237
information revolution 67, 80
Information Society 10, 118
innovation 14, 21, 27–8, 31–2, 35–7, 40, 47–9, 53, 60, 63, 67–9, 72, 77, 80, 93, 95, 101, 106–9, 111–12, 115, 128, 136, 152, 156–8, 180–4, 188, 207, 212, 217, 248–50, 253–4, 260
institutional reflexivity 106
instituts universitaires de technologie 86
instrumentation 106, 111, 119, 124, 145, 185
intellectual property 171, 176, 238
interdisciplinarity 89, 228
internal and external selection criteria 122–3
internal market 22, 24, 71
internationalization 7, 52
Internet 11, 25, 54, 221, 261
irrationality 83, 198, 203, 207
Israel 8

Jacob, François 35
Japan 52, 185–7
Jasanoff, Sheila 134
job creation 75–6

Katz, J. Sylvan 3
Kennedy, P. 12
Keynes, John Maynard 9, 33, 112
Knorr Cetina, Karin 97–100, 103, 114, 125
knowledge industries 15
knowledge matrix 118
Knowledge Society 10, 12, 15–16, 30, 56, 74, 82, 89–90
knowledge workers 15–16
knowledgeability 80–1, 84
Kruckemeyer, Kenneth 138
Kuhn, Thomas 146, 190, 192, 195, 199

laboratories 62, 67, 70, 72–3, 78, 99–100, 114, 124, 155, 157, 167, 183, 188, 197, 203
labour market 83
Latour, Bruno 2–4, 19, 54, 66, 140, 203
lay people 10, 57, 103, 133, 156, 186, 190, 211, 229, 234, 240
learning organizations 74, 89
Leavis, F. R. 81
liberal democracy 5, 23, 40, 63, 68, 119, 203–4, 209, 212, 222, 226
life sciences 41, 54, 60, 125, 165
life-style 40, 46, 56, 62, 91, 118, 182, 201, 220, 251
Luhmann, Niklas 32, 34, 201–2, 236–7

MacKenzie, Donald 42
management consultancy 15, 46
Manhattan Project 129
manufacturing 105
market 4–5, 7, 11–12, 17, 21–6, 28–32, 38, 40, 67–8, 73, 82, 85–6, 88, 97, 107, 109–10, 118, 122, 126, 129, 151–2, 154, 166, 180, 183–4, 186, 192, 203, 206–7, 209–11, 215, 220, 222, 245–7, 253
marketing 107, 154
mass consumption 6, 117
mass higher education 81–4, 90, 93, 104–5, 204
mass production 6, 117, 136
materialization 11, 28
Max Planck Society 85–7
mechanical objectivity 64
media 9, 16, 21, 23, 25–6, 32, 38, 43, 68, 74, 88, 111, 155–6, 164, 166, 189–90, 206, 209–10, 212, 221, 224–5, 241, 246, 248, 261
Medical Research Council (MRC) 148–9
medicine 36, 40, 61, 80, 139–40, 142, 144, 148, 216
Megascience Forum 126
Mermin, David 195, 250
Merton, Robert 59, 241
meta-discourse 19

Middlemarch 138
military 6, 8, 60, 67, 79, 107, 124, 129, 135–6, 150–1, 206
ministries 74, 155–6
missile testing 42–3
Mode 1 15, 47, 54, 95, 167, 175, 206
Mode 2 3–4, 15–18, 22, 29, 47, 58, 74, 77, 80, 82, 84, 88–90, 92, 95, 113, 134, 141, 144, 147, 167, 174–7, 242, 248, 260
Mode-1 science 11–12, 15–16, 28, 80–2, 88–90, 96, 106, 112, 114, 134, 141, 149, 158–9, 176, 220, 242
Mode-2 contextualization 19
Mode-2 institutions 79
Mode-2 knowledge production 11–12, 15–16, 18–22, 28–30, 32–3, 47, 50, 54, 69, 74, 89–90, 93, 96, 114, 147, 159, 174, 176, 186, 203, 223, 226–8, 233, 239–40, 242, 252
Mode-2 objects 97, 126, 144, 147–55
Mode-2 science 1, 4, 10, 19, 158, 245
Mode-2 society 4, 10–11, 19, 29, 33–4, 47, 50, 54, 68, 89, 91, 94, 96, 102–3, 110, 112–13, 115–16, 118, 130, 145, 147, 159, 179–80, 197, 203–4, 206, 209, 211, 214–16, 226–8, 230, 233, 235–6, 239–40, 242, 245, 248–9, 252–3
Mode-2 university 91, 93
modelling 7, 106, 133, 165, 213
modernity 1, 3–6, 14, 21, 28, 30, 32, 47–8, 50, 94, 106, 135, 163, 184, 190–1, 195, 201, 203–6, 211, 216–22, 243, 245
modernization 5–6, 14, 29, 44, 47, 63, 94, 117, 180, 184, 216–17, 219–20, 229
molecular biology 54, 97–100, 103, 114, 141
multi-disciplinary research 76, 106
multinational corporations 41, 70–1
multiple authorship 98, 100, 125
muscular dystrophy 140–1
museums 28, 241
Musil, Robert 51

nanotechnology 129
NASA 128
National Human Genome Research Institute (US) 238
National Institutes of Health (US) 61
national identity 117
national research programmes 70, 121, 127–30, 133, 141, 148
nationalized industry 70–1
nation-state 13, 22–4, 26, 31, 63, 118–19, 128, 211, 217–20
natural environment 6–7, 53, 56, 72, 133, 187, 230–1
natural philosophy 55, 243
natural sciences 63, 72, 81, 117, 164
nature 5, 39, 62–3, 103, 111, 122–3, 159, 168, 173, 178, 184–5, 193–5, 202–4, 211, 220, 226, 235, 243
Nature 54, 60, 76
Nelkin, Dorothy 212
Netherlands, the 76, 85
Neurath, Otto 178
New Production of Knowledge, The 1, 3–4, 19, 22, 30, 42, 47, 50, 104, 144, 174
New Public Management 24
New Zealand 73
non-governmental organizations (NGOs) 69, 132–3, 213
North America 6, 79, 162
Notgemeinschaft der Deutschen Wissenschaft 59
Nowotny, Helga 155
nuclear power 2, 7, 34, 124
nuclear proliferation 13
NWO 76

objectivity 2, 18–19, 42, 54–5, 63–4, 84, 116, 169, 178, 197–9, 208, 210–12, 230, 232, 235
objectivity trap 55, 116
oil crisis 6–7, 9, 12
Organization for Economic Co-operation and Development (OECD) 126

Panofsky, Erwin 243
particle physics 121, 124–7, 147, 193
partnership 92
patents 60, 74
patients 10, 15, 121, 139–42, 160–1, 205, 213
peer review 46, 60–1, 76–7, 115, 169, 209, 212
performance indicators 24
pharmaceutical industry 40, 142
'pidgin' language 146, 175
Plato 203–4
Poland 234
politics and politicians 2, 4–6, 8, 17, 21–3, 26, 28, 38, 49, 59, 63, 67, 80, 86, 118–19, 122, 124, 126–7, 130, 132–4, 137–9, 148, 155–6, 164, 179, 183, 206, 209, 214–15, 222–4, 233
polytechnics 86
Popper, Karl 159, 194, 199
Popperian falsifiability 48
popularization of science 74, 90, 189, 259
Porter, Theodore 219
positivism 104
post-Fordism 11
post-industrialism 11–12
post-materialism 13
post-modern condition 9–10
post-modernism 9, 11, 17
post-war period 79, 105, 136; *see also* after 1945
Power, Michael 45, 239
President's Council of Economic Advisers 67
pressure groups 115, 152
privatization 22, 25, 73, 207, 229
productivity 11, 15, 56, 128, 207
professional ethics 24
professional identity 104, 166, 177
professionalization of science 46, 49, 61, 79, 193
Prometheus 27
public controversy and contestation 83–4, 148, 182, 188, 196, 201–4, 208–11, 217, 222, 225, 231, 258, 261
public opinion 110, 132, 183, 203, 208, 210, 213, 221
public relations 111, 139, 197
public sector 24, 113
public-service ethic 24
public understanding of science 197, 240, 260
pure research 80, 108, 113, 122, 184, 220

quality assurance 92
quality control 17, 46, 60, 115, 174, 219, 227

R&D 4, 71–2, 79–80, 88, 95, 107, 136, 151, 166, 184
Rabinow, Paul 183
race 11, 137
rationality 21–2, 24–5, 30–3, 37–8, 47, 63, 93, 183, 190, 216, 218
reductionism 19, 49, 167, 199, 243
reflexive modernization 17, 19, 28–9, 44
reflexivity 44–8, 51, 89–90, 106, 108–9, 145, 209, 228, 254
regulation 23, 25, 61, 69, 74, 151–2, 202, 210, 212, 218
Reich, R. 12
relativism 4, 84, 197, 207, 259
reliable knowledge 42, 84, 154, 168–79, 197–9, 201, 210–11, 232–3, 240, 246, 248, 258, 261
religion 50, 122, 185–6, 197, 253
Renaissance 242–4
research agendas 56, 122, 124, 127–8, 131, 133–4, 141, 161, 163–4, 213–14, 241
research councils 47, 50, 56, 69–70, 73, 75–8, 87, 95–6, 128, 156, 158, 166, 241
research institutes 15
retinitis pigmentosa 141
risk 33–5, 39, 42, 45, 72, 112–13, 119, 145, 181, 201–2, 208, 210–12, 218
Risk Society 10, 14–15, 30, 34, 47
Risk Society 14
Rorty, Richard 18
Rostand, Jean 190

Royal Society 55, 63
Rubbia, Carlo 125
Russia 138

SAGE air defence system 135
Schaffer, Simon 52, 231
schools 81, 86, 241
Schumpeter, Josef 48
Science 2
science policy 3, 11, 22, 67, 107, 109–10, 122–3, 132, 156, 207, 252
scientific authority 29, 51, 117, 180, 198, 217, 220–1, 224–7, 247–9, 257, 261
scientific autonomy 58, 115, 119, 169, 195, 204, 210, 230, 232
scientific community 2–3, 49, 62, 84, 89–90, 103–4, 110, 113, 124–5, 128, 130, 177, 190, 194–6, 198, 246, 250, 258, 261
scientific creativity 115, 158, 197, 232
scientific elite 3, 47, 104, 112–13, 132
scientific literacy 81, 83
scientific method 4, 54, 106
scientific publications 87
scientification of society 3, 90, 191, 207
Scotland 183; *see also* Britain; United Kingdom
selective retention 114–15, 121, 131, 143, 248
self-control 45–6
self-organization 44, 48, 118, 158, 197, 236, 253, 261
Sen, Amartya 240
Shakespeare 27, 138
Shelley, Mary 189
simulation techniques 106, 111, 145
skills 42, 45, 69, 72, 99–100, 105, 113, 118, 153, 176, 182, 215
Snow, C. P. 81
social class 11, 43, 85, 87, 181, 217
social control 24, 45–6, 58
social distance 42–4, 122
social engineering 5, 9
social forecasting 30
social knowledge 63
social mapping 14
social policy 23
social reform 8
social sciences 6, 63, 72, 102, 105–6, 117, 161–2, 164, 189, 195, 207, 235
social status 83–4, 217, 219
social transformation 3, 10, 16
socially distributed expertise 35, 104–5, 214–29, 247–8, 254, 261
socially robust knowledge 117, 167–8, 178, 197–9, 201, 210–11, 227, 232–3, 240, 246–8, 258, 261
socio-economic development 10, 56, 181
Socrates 203
South Africa 8
Soviet Union 124, 138
space flight 57, 128
state 7, 12, 21–5, 27–31, 47, 67–71, 87–8, 107, 110, 113, 117–18, 123, 126, 128, 143, 149, 166, 191–2, 206, 211, 218–19, 221, 226, 229, 235, 245, 247
Stehr, Nico 82
Stevenson-Wydler Technology Innovation Act 60
Strathern, Marilyn 45, 115, 229
strong contextualization 78, 97, 120–1, 129, 131–43, 154, 167–8
Structure of Scientific Revolutions, The 192
super-conducting super-collider 126, 194
super-organisms 97, 125
supply chain 72, 116, 118, 133–4, 180
sustainability 7, 53, 72, 213
Sweden 85
Switzerland 75–6, 155, 209
systems theory 44, 202, 237

Taylor, Frederick 138
teaching 35, 81, 83, 86–8, 90–1, 94
Technical High Schools 108
technocracy 51
technology foresight exercises 70, 182, 217, 233, 241, 254–5
think-tanks 15

Third Way 10
Third World 40
Three Gorges Project (China) 138
Thucydides 161
tourism 28
trading zones 24, 143–7, 175, 177, 252
training 68, 83, 87, 90, 92, 94, 169, 173, 175, 197, 219
transaction spaces 97, 103, 141, 143–7
transdisciplinarity 89, 223
transgressivity 21, 225
Traweek, Sharon 125
Turney, Jon 188–9
unemployment 14, 251

United Kingdom 73, 81, 85, 148–9, 223–4; *see also* Britain; Scotland
United Nations 213
United States 8, 26, 59–61, 67, 73, 76, 88, 126, 135, 159, 194, 228
university 15–16, 38, 50, 53, 59–60, 67, 69–70, 73, 75, 77–88, 90–6, 108–9, 122, 128–30, 132, 140–1, 166, 175, 179, 204, 207, 212, 219, 236, 238–9, 241–2, 247, 251, 255; *see also* higher education
urbanization 117–18, 216
utilitarianism 31
Utterback, J. 153–4

value systems 122, 166
Van Duinen, Reinder 77
Vienna Circle 234
Vietnam 9
virtual communities 33
virtual products 11, 25

war 63, 105, 218
wars on cancer 129
weak contextualization 77, 97, 120–31, 143, 154, 167–8
wealth creation 31–2, 53, 56, 109, 166, 175, 180–1, 241
Weber, Max 24, 46, 219
Weinberg, Alvin 122–4, 126
Weinberg, Steven 192–5, 260
welfare state 5, 7, 9, 12–13, 31–2, 117, 217–18, 251, 253
West, the 5–8, 11, 23, 40, 51–2, 63, 68, 70, 94, 102, 126–7, 138, 181, 185–7, 205, 209, 222, 226, 234, 253
Wittgenstein 156
women's studies 140
work 33, 118, 125, 185, 237, 251–2
Wynne, Brian 208

Ziman, John 4, 169–75, 198